Mark Carwardine's
Guide to
WhaleWatching

For Jessica and Zöe, with love

First published in 2003 by New Holland Publishers (UK) Ltd
London • Cape Town • Sydney • Auckland

10 9 8 7 6 5 4 3 2 1

Garfield House, 86-88 Edgware Road, London W2 2EA,
United Kingdom
www.newhollandpublishers.com

80 McKenzie Street, Cape Town 8001, South Africa

Level 1/Unit 4, 14 Aquatic Drive, Frenchs Forest, NSW 2086, Australia

218 Lake Road, Northcote, Auckland, New Zealand

Please note, all websites were correct at time of going to press.

ISBN 1 84330 059 1

Publishing Manager: Jo Hemmings
Project Editor: Lorna Sharrock
Copy Editor: Sylvia Sullivan
Editorial Assistant: Gareth Jones
Designer and cover design: Katie Benge at Design Revolution
Cartographer: Bill Smuts
Index: Janet Dudley
Production: Joan Woodroffe

Front cover: Humpback Whale, lobtailing
Page 3: Fraser's Dolphin, porpoising
Page 4: Humpback Whale, breaching; Humpback Whale, underwater
Page 12: Humpback Whale, lobtailing
Page 22: Humpback Whale underwater
Page 60: Short-beaked Common Dolphins, from the *Pride of Bilbao*

Reproduction by Pica Digital Pte Ltd, Singapore
Printed and bound by Kyodo Printing Co (Singapore) Pte Ltd

WDCS
Whale and Dolphin Conservation Society

Mark Carwardine's
Guide to
WhaleWatching

Britain and Europe
• Where to Go • What to See

NEW HOLLAND

CONTENTS

 # The Whale and Dolphin Conservation Society

(WDCS) is the global voice for cetaceans (whales, dolphins and porpoises) and their habitats. Founded in 1987, WDCS has its headquarters in the UK, but has offices in several different countries, as well as an international network of consultants, so our reach truly is global. WDCS is unique in combining welfare, conservation and campaign issues relating to cetaceans.

We have over a decade of experience in promoting sustainable whalewatching activities in different parts of the world, working with whale-watch operators, the tourism industry and the public to ensure that the whale-watch experience on offer is educational and enjoyable and, most importantly, respects the whales and dolphins and their habitat. WDCS believes that offering good-quality whalewatching is a sound ethical and economic means for coastal communities to make a livelihood.

Mark Carwardine has been a good friend and valued consultant to WDCS for many years. He has introduced so many people to the thrill of watching whales and dolphins in their natural habitat, through his books, television and radio broadcasts and by being a truly inspirational leader of whale and dolphin watching adventures all over the world. Meanwhile, enjoy browsing through this wonderful guide to our whales and dolphins – you will soon be planning your next whale-watch trip!

To find out more about our work, please visit our website at www.wdcs.org

 ## The Wildlife Trusts

partnership is the UK's leading voluntary organization working, since 1912, in all areas of nature conservation. We are fortunate to have the support of over 382,000 members – people who care about British wildlife.

We protect wildlife for the future by managing almost 2,500 nature reserves across the UK, ranging from wetlands and peat bogs, to heaths, coastal habitats, woodlands and wildflower meadows.

We are increasing our efforts to protect Britain's marine environment. The Wildlife Trusts partnership works with Government, companies, including those within the fishing industry, other conservation bodies and the general public, to raise awareness of the importance of threatened habitats and the need to protect them. Our marine team has been leading on trials for an electronic device, which attaches to nets and logs the echo-location clicks of dolphins and porpoises. These will drastically reduce the numbers of by-catch. And at time of writing our Basking Shark expert Colin Speedie is undertaking a major three-month survey of these gentle giants off our west coast.

We encourage people to 'do their bit' for wildlife. Help us to protect wildlife for the future and become a member today! Please phone The Wildlife Trusts on 0870 0367711 or log on to www.wildlifetrusts.org for further information. The Wildlife Trusts is a registered charity (number 207238).

How to Use this Book

The book is divided into three main sections:

1 **Whales and Whalewatching.** An introduction to the natural history and biology of whales, dolphins and porpoises as well as information on the techniques and responsibilities of whalewatching.

2 **Identifying Europe's Whales.** An identification guide to all 36 species of whales, dolphins and porpoises found in Europe. Some have been recorded just a handful of times, or occur in Europe at the extreme edge of their range, but the majority are resident or regular visitors in the region. Each entry includes a full-colour illustration, an introduction to the species, key facts at a glance (alternative names, scientific name, size, diet, behaviour and distribution), an identification checklist, a European distribution map (showing the range of the species during a typical year) and small images of its dive sequence and any other interesting surface behaviour.

3 **Where to Watch Whales in Europe.** A detailed directory of whalewatching opportunities in 19 different countries, independent territories and regions in Europe. It is possible to see cetaceans elsewhere in the region, of course, but this directory includes all the hotspots with outstanding opportunities and organized or commercial whalewatching. Each entry includes a full-colour site map (showing all the places mentioned in the text), practical details (the main species likely to be seen, the key locations, types of tours, what time of year to go and contact details for whalewatch operators) and a detailed description of what to expect and the main whalewatching opportunities.

Map Key			
Road bridge	————	City/large town	Havneby ●
International border		Ferry route	————
Arctic Circle	– – – -	Mountain	▲

Cetacean Anatomy

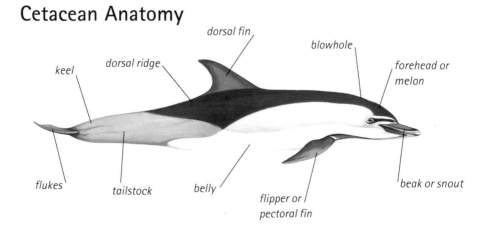

dorsal fin

blowhole

keel

dorsal ridge

forehead or melon

flukes

tailstock

belly

flipper or pectoral fin

beak or snout

Introduction

Imagine watching Minke Whales in the golden light of the midnight sun, snorkelling with a wild Bottlenose Dolphin, or following a family pod of Killer Whales hunting Herring in a remote Arctic fjord. These are just a few of the world-class whale- and dolphinwatching experiences now available in different parts of Europe.

Worldwide, whalewatching has grown from humble beginnings in the mid-1950s, when people first began to take an interest in grey whales migrating up and down the coast of California, to today's US $1 billion industry involving nearly

90 different countries and overseas territories. No fewer than nine million people now join commercial whalewatching trips around the world every year.

Europe is not yet as well known for its whales, dolphins and porpoises as North America, New Zealand, Antarctica and some other places, but it is home to a surprisingly wide variety of species. Blues, Fins, Minkes, Humpbacks, Killer Whales, Sperm Whales and Pilot Whales are some of the larger species that can be seen, as well as an interesting variety of dolphins, and Harbour Porpoises. It is one of the best places to look for Northern Bottlenose Whales and Cuvier's Beaked Whales and some tours offer Arctic specialities such as Beluga and Narwhal.

It is even possible to choose *how* to watch them: from the air, from the shore, or from a host of different vessels, including yachts, rubber inflatables, motor cruisers, research boats, kayaks and huge ocean-going ships. There are tours to suit every taste, from comfortable half- or full-day excursions to adventurous two- or three-week expeditions, and many of them are genuinely world-class, combining recreation, education, science, conservation and good business sense to provide a high-quality service. Perhaps, then, it is not surprising that Europe is rapidly gaining a reputation as a whalewatching hotspot.

The first commercial whalewatch trips here took place in Gibraltar, in 1980, when one man began to take people to see three species of local dolphins. By the mid-1980s, dolphinwatching trips were being offered in France, Germany, Britain and Ireland. Today, no fewer than 19 European countries and territories, and nearly 70 different communities, are involved in whalewatching of one kind or another. From Portugal to Croatia and from Greece to Greenland, they attract huge numbers of whalewatchers every year.

According to studies by Erich Hoyt, who works for IFAW and WDCS, Italy, Spain and Iceland are among the six fastest-growing whalewatch industries in the world. In fact, Iceland is experiencing one of the highest growth rates ever recorded – its industry expanded by an average of 250.9 per cent every year from the mid- to late-1990s. In 2001, an impressive 60,550 people went whale-watching from nine different communities around the

Opposite: *A Minke Whale* (Balaenoptera acutorostrata) *surfacing in the golden light of the midnight sun, Skjálfandi Bay, northern Iceland. Notice the distinctive body and fin shape.*

country, which is equal to more than one out of every eight visitors. Iceland is widely considered one of the newest and most exciting areas for whalewatching in the world, not least because it stopped hunting whales only in the late 1980s; visitor surveys suggest that the whalewatch industry might suffer if whaling is ever resumed and this could prevent it happening.

There is also huge potential in the Mediterranean Sea, where whalewatching has been given a new lease of life with the recent designation of the Mediterranean Cetacean Sanctuary by Italy, France and Monaco. Meanwhile, whalewatching in the Canary Islands has grown to such an extent that it has become one of the top three whalewatch destinations in the world. The Canaries can now claim more than a million whalewatchers every year and only the United States and Canada are able to do the same.

Some of the most unlikely (and most rewarding) whalewatching takes place from the P&O Portsmouth and Brittany Ferries passenger cruise-ferries operating between the UK and Spain. They travel across the Bay of Biscay, which, in recent years, has earned a well-deserved reputation as one of the most exciting, accessible and affordable places in Europe to watch an impressive variety of cetaceans. More than a dozen species are seen regularly from the observation decks and a number of others, including rarely seen beaked whales, have been recorded as well.

Aims of this book

The aim of this book is to provide a comprehensive guide to shore-based and boat-based whalewatching in Europe. It highlights the most exciting and reliable places to watch

Swimming with whales and dolphins

There are some (although relatively few) opportunities in Europe for swimming with whales and dolphins. Be aware, however, that the cetaceans do not necessarily share the excitement and pleasure we derive from this activity and, in some circumstances, may find it threatening. In-water encounters can also be unsafe for human swimmers.

whales, dolphins and porpoises and helps you to select a whalewatch trip according to your main interests. It explains where and when to go, what to expect when you get there and who to contact in order to book a tour. The comprehensive identification section introduces the species you are most likely to encounter and explains how best to watch them.

So, if you feel like sharing a day with a group of Blue Whales against a spectacular backdrop of snow-capped mountains, observing Harbour Porpoises from the comfort of your hotel bed, or even searching for rare beaked whales from the deck of a cruise-ferry, then look no further than Europe. It really does offer some of the most memorable and exciting whalewatching in the world.

Below: *Whalewatchers enjoy a close encounter with a Short-finned Pilot Whale (*Globicephala macrorhynchus*) off the coast of Tenerife, in the Canary Islands.*

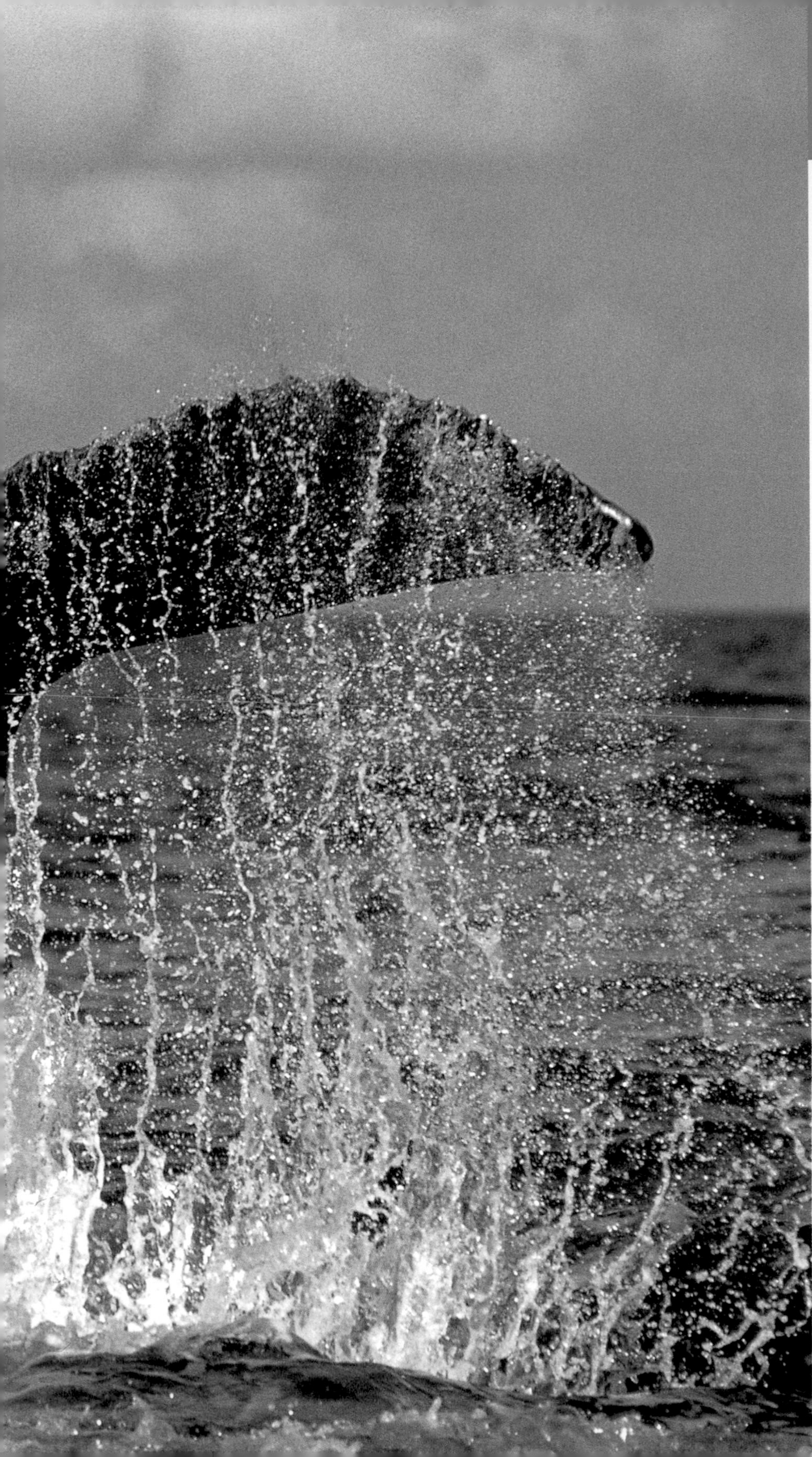

1 WHALES AND WHALEWATCHING

Understanding Whales, Dolphins and Porpoises

Scientists on a fossil-hunting expedition to Pakistan, in the late 1970s, made an extraordinary discovery. They found part of the skull of a dog-sized mammal protruding from a 50-million-year-old boulder. Named *Pakicetus inachus*, it was the oldest and most primitive whale known to science. Its owner had descended from land mammals and, although it had an elongated whale-like body, it probably possessed both front and hind legs and was likely to have been amphibious. It was classified in a group of animals known as the ancient whales, or archaeocetes, which are believed to have shared the same ancestors as modern whales.

Reproduction

There are almost as many different reproductive strategies as there are species of cetacean. The age of sexual maturity varies greatly from species to species: at one extreme, male Sperm Whales are not sexually mature until they reach their late teens or early twenties; at the other extreme, female Harbour Porpoises reach sexual maturity at about three years old and begin to breed during the following year.

There are no records of cetaceans being monogamous: it has never been known for one male to mate exclusively with one female. The gestation period ranges from about 10 months to at least 17 months. A single calf is born underwater (twins are extremely rare) and it is able to swim, albeit a little awkwardly, almost straight away. Suckling takes place underwater, and the mother literally squirts her rich, fat-laden milk into the young animal's mouth. Weaning is usually gradual, with a period of overlap when the calf is drinking milk and eating solid food at the same time. It stays with its mother for weeks, months or even years afterwards, depending on the species.

The maximum lifespan is unknown for most cetaceans, but probably ranges from about eight years in the Harbour Porpoise to as long as 200 years in the Bowhead Whale.

No-one knows exactly how many species of whales, dolphins and porpoises are alive today, but most experts quote a figure of 85 or more. New ones are still being discovered, and the latest genetic research is revealing that some that we previously thought to be single species should actually be split into two or more.

Whales, dolphins and porpoises come in a variety of shapes and sizes, ranging from the tiny Hector's Dolphin, just over 1m (3ft) long, to the enormous Blue Whale, which is almost as long as a Boeing 737. Known collectively as cetaceans, they form the largest of the three main groups of marine mammals surviving today; the others are the seals, sealions and walrus, or pinnipeds, and the manatees and Dugong, or sirenians.

The two groups of cetaceans
Modern cetaceans fall into two distinct groups: the toothed whales, or odontocetes, which possess teeth; and the baleen whales, or mysticetes, which do not. The baleen whales are unique in having hundreds of furry, comb-like

Below: Humpback Whales (Megaptera novaeangliae) are among the most studied of all the world's whales and are particularly popular with whalewatchers, because of their spectacular surface behaviour.

'baleen plates' or 'whalebones' hanging down from their upper jaws, which form a bristly sieve for filtering small prey animals out of the seawater.

Adaptations to underwater life

Whales are so streamlined, and so well adapted to life underwater, that they resemble sharks and some other large fishes. But appearances can be deceptive and the two groups are entirely unrelated; they have simply adapted to similar conditions. Fishes are cold-blooded, use their gills to extract all the oxygen they need directly from the water, and normally lay eggs or give birth to young that can feed themselves. Whales, dolphins and porpoises are mammals, like us, and so are warm-blooded, breathe air with lungs, and give birth to young that feed on their mother's milk for the first weeks or months of life. Most adult cetaceans feed on fish, squid or crustaceans, although some will take a wide variety of other prey as well.

Below: *No two Atlantic Spotted Dolphins (*Stenella frontalis*) look alike: the amount of spotting varies greatly between individuals and from region to region.*

Unlike most other mammals, whales, dolphins and porpoises never fall into a deep sleep, and do not seem to have regular sleeping patterns linked to night and day. They do sleep, of course, but not in the way we might expect. They control their breathing consciously, so have to be awake and able to think in order to take regular breaths. If they were to fall into a deep sleep, which is a naturally-occurring state of unconsciousness, they would drown. Instead of long periods of deep sleep, their solution is to swim along slowly, or lie just below the surface, and take short 'cat-naps'. They are believed to go into a semi-conscious state by switching off one half of the brain at a time. Then they swap sides to ensure that both halves are fully rested.

They find their way around their underwater world in ways that are well beyond the scope of our own senses. Most of them retain the senses of sight, touch and taste, but they rely mainly on hearing. This works at night, as well as during the day, and in poor visibility where eyesight would be useless. Although cetaceans have lost the distinctive outer ear flaps typical of most other mammals, they do have ears in the form of tiny holes in the skin behind their eyes and many species can probably also receive sounds through their lower jaws. They also have a remarkable sensory system, called echolocation, which enables them to build up a 'sound picture' of their underwater surroundings. Basically, they transmit ultrasonic clicks into the water and then monitor and interpret any echoes that bounce back. It is actually a form of sonar, but the system used by humans is little more than a crude imitation by comparison. Toothed whales are the real echolocation experts; baleen whales may also be able to echolocate, but this has never been proven and their system is probably much less well developed.

Extra sense

Many cetacean species are also believed to have an extra sense. This sense is known as geomagnetism, and it is thought that it enables them to 'read' the Earth's magnetic field like an invisible map when they are navigating over large distances.

Current threats

To the best of our knowledge, no whale, dolphin or porpoise has become extinct in modern times. A frightening number, however, are in serious trouble, and others have all but disappeared from many of their former haunts.

Since its early beginnings nearly 1,000 years ago, commercial whaling has had a long and chequered history. The sheer variety of products derived from whales is quite astonishing and, for several centuries, these products touched almost every aspect of daily life in Europe and North America. In an age before petroleum or plastics, whales provided valuable raw materials for thousands of everyday products, from soap and candles to whips and corsets. The slaughter reached its worst excesses around the middle of the twentieth century and, one by one, the great whales were hunted almost to the point of extinction. Literally millions of them were killed around the world and, today, we are left with no more than the tattered remains of their earlier populations. Incredibly, we still have not learned the lessons of the past and commercial whaling continues as Norway and Japan persist in hunting whales in the North Atlantic, the North Pacific and the Antarctic.

Since the 1950s, the staggering growth of many modern fisheries and the introduction of increasingly destructive fishing methods have also spelt disaster for whales, dolphins and porpoises around the world. Hundreds of thousands of them – possibly even millions – die slow, lingering deaths in fishing nets every year. Many more could be threatened by the sheer scale of modern fisheries, which over-exploit fish stocks with scant regard for the future health of the world's oceans.

Some experts predict that pollution could become the most serious threat to whales, dolphins and porpoises in the future. It is a silent, insidious and widespread killer and yet ever-increasing quantities of industrial waste, agricultural chemicals, radioactive discharges, untreated sewage, oil, modern plastic debris and a wide variety of other pollutants are dumped directly into the seas and oceans every day. Underwater noise pollution, caused by a variety of human activities from coastal development and seismic testing to speedboats and heavy shipping, is also very worrying since whales, dolphins and porpoises rely on sound for many of their day-to-day activities. Meanwhile, habitat degradation and disturbance such as coastal and riverbank development, land reclamation, deep-sea dumping, oil, gas and mineral exploration, commercial fish farming, boat traffic, and the effects of land-based activities such as deforestation and river damming all take their toll.

How to Watch Whales, Dolphins and Porpoises

It is possible to see whales, dolphins and porpoises almost anywhere in the world. But good planning is essential, because many species tend to occur only in particular areas or at certain times of the year. Even then, they can be difficult to find, and patience is an important prerequisite to any whalewatch trip.

With a large whale, the first clue is often its blow or spout. This is more visible in some weather conditions than others, but it can be surprisingly distinctive. It may look like a flash of white or a more gradual puff of smoke.

Below: *One species every whalewatcher wants to see is the Killer Whale (Orcinus orca), or Orca, which is actually a member of the dolphin family. This one has quite a visible blow – the cloud of water vapour that it is exhaling.*

19

Alternatively, you may catch a glimpse of the head and back of the whale as it breaks the surface. Splashes are also good clues and can be caused by a large whale breaching, flipper-slapping or lobtailing – or by dolphins. The presence of birds can often be a tell-tale sign, as well, particularly if they seem to be feeding in one particular area.

Identification

Identifying whales, dolphins and porpoises at sea can be quite challenging. In fact, it can be so difficult that even the world's experts are unable to identify every species they encounter: on most official surveys, at least some sightings have to be logged as 'unidentified'. However, it is quite possible for anyone to recognize the relatively common and distinctive species and, eventually, many of the more unusual ones as well.

The easiest way to identify them is to use a relatively simple process of elimination. This involves running through a mental checklist of key features every time a new animal is encountered at sea (see **Checklist of identification features**). It is almost never possible to use all these features together, and one alone is rarely enough for a positive identification, but the best approach is to gather information on as many as possible before drawing any firm conclusions.

Checklist of identification features

- Geographical location and habitat
- Approximate size
- Unusual features
- Size, shape and position of the dorsal fin
- Length, colour, shape and position of the pectoral fins
- Body shape
- Presence or absence of a prominent beak
- Unusual colours or markings
- Fluke shape and markings
  (most useful when larger whales raise their tails at the beginning of a dive)
- Height, shape and visibility of the blow
- Dive sequence (noting details such as the angle at which the head breaks the surface, how strongly the tailstock is arched, etc.)
- Interesting surface behaviour
- Number of individuals travelling together

It is often tempting to guess the identification of an unusual whale, dolphin or porpoise that you have not seen very clearly. But this is a mistake. Apart from the fact that it is bad science, it does little to improve your identification skills. Working hard at identification – and then enjoying the satisfaction of knowing that an animal has been identified correctly – is what makes a real expert in the long term.

Do not disturb!

Finally, a word of caution. Whalewatching should be an eyes-on-hands-off activity. Most whalewatch operators take their responsibility to cause as little disturbance as possible very seriously. They put the welfare of the whales first, manoeuvring their boats carefully, slowly, and not too closely, and then leaving before the animals show signs of distress. They also have knowledgeable naturalists on board to keep everyone well informed, provide free places for biologists to do urgently needed research, and help to raise money for whale conservation. However, some operators are not so careful. They cause unnecessary disturbance, or even injure the whales, and make no contribution to education, research or conservation. Therefore, it is crucial to take the time and trouble to choose the best and most responsible operators before booking a trip.

Left: *A homemade directional hydrophone, or underwater microphone, is used to track Sperm Whales (*Physeter macrocephalus*) underwater off the coast of Faial, in the Azores.*

2 IDENTIFYING EUROPE'S WHALES

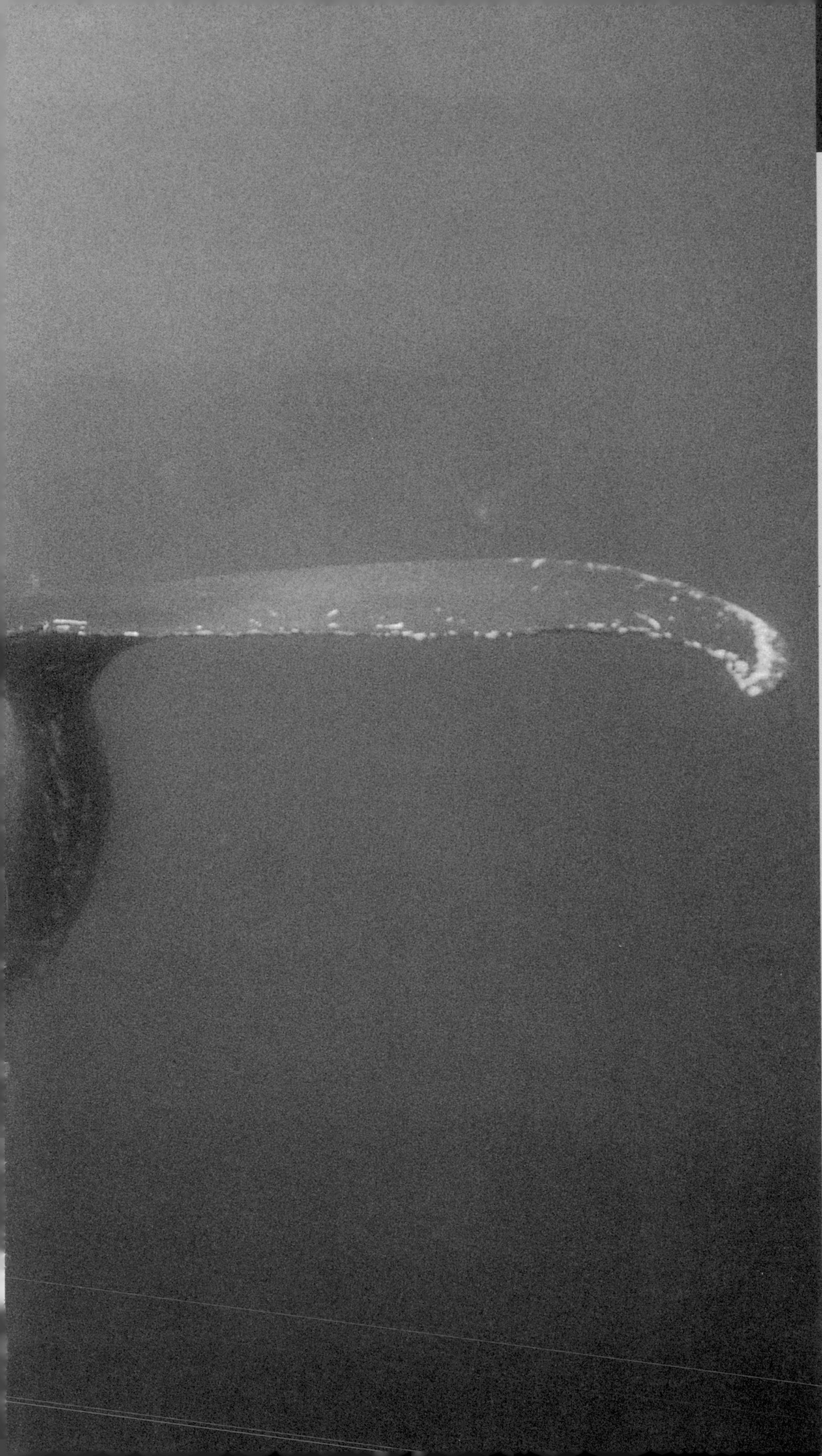

Bowhead Whale

Identification

- V-shaped blow
- no dorsal fin
- dark body colour
- irregular white patch on chin
- two distinct humps in profile
- large head with arching mouth
- no callosities or barnacles
- often raises flukes before diving

Named for its enormous and distinctive bow-shaped skull, the Bowhead is the only large whale that lives exclusively in the Arctic. With a layer of blubber up to 70cm (28in) thick and an ability to create its own breathing holes by breaking through the ice, it is well adapted to life in its freezing home. When lying at the surface, most adults show two distinct humps in profile, making them resemble the supposed shape of the Loch Ness Monster; the triangular hump in front is the head, the depression is the neck, and the rounded hump in the rear is the back. The Bowhead has the longest baleen plates of any whale: many have been recorded over 3m (9½ft) long and there is a disputed claim of one measuring 5.8m (19ft). Heavily hunted by commercial whalers for several centuries, a few Bowheads are still being taken each year by the Inuit in the USA and Canada, and by other native people.

At a glance

Alternative names: Greenland Right Whale, Greenland Whale, Arctic Right Whale, Arctic Whale, Great Polar Whale.
Scientific name: *Balaena mysticetus*.
Adult size: 14–18m (46–59ft); 60–100 tonnes; females larger than males.
Diet: mainly krill, copepods and other small- and medium-sized crustaceans; may also feed on other invertebrates on or close to seabed.
Behaviour: slow swimmer but may breach, lobtail, flipper-slap and spyhop.
Distribution: cold Arctic and sub-Arctic waters, rarely far from pack ice; normally migrates to the high Arctic in summer, but retreats southwards with advancing ice edge in winter; very rare in Europe.

Distribution map

Dive sequence

Fluking

Spyhopping

Breaching

Northern Right Whale

With no dorsal fin, a dark rotund body, and an enormous head covered in strange, hardened patches of skin, the Northern Right Whale is relatively easy to identify. It is virtually identical to another species known as the Southern Right Whale, which, as its name suggests, lives in the southern hemisphere. The patches of skin are known as callosities and, although their function is still a mystery, each whale has a different arrangement on its head, and scientists are able to use them to tell individuals apart. Northern Right Whales were probably the first whales to be hunted commercially: Basque whalers were killing them in the Bay of Biscay as early as the twelfth century. They were popular targets for commercial whalers for many centuries and, although they have been protected since the mid-1930s, show no sign of recovery. The vast majority of the 300-odd survivors live in the western North Atlantic, with no more than a handful in Europe and the eastern North Pacific. The Northern Right Whale is probably closer to extinction than any other large whale – and may never recover.

At a glance
Alternative names: Black Right Whale, Biscayan Right Whale.
Scientific name: *Eubalaena glacialis.*
Adult size: 11–18m (36–59ft); 30–80 tonnes; females slightly larger than males.
Diet: tiny crustaceans, called copepods; sometimes also shrimp-like krill.
Behaviour: slow swimmer but surprisingly acrobatic and frequently breaches, flipper-slaps and lobtails; can be very inquisitive.
Distribution: main population in the western N. Atlantic, but occasional sightings in Europe and the eastern N. Pacific.

Identification

- V-shaped blow
- dark, rotund body
- broad back with no dorsal fin
- large head covered in callosities
- strongly arched mouthline
- large paddle-shaped flippers
- often raises flukes before diving

Distribution map

Historic range

Dive sequence

Fluking

Sailing

Spyhopping

Breaching

Minke Whale

Identification

- low, indistinct blow
- sharply pointed snout
- snout breaks surface first
- relatively large, falcate fin
- white bands on flippers in some populations
- single longitudinal ridge on head
- tailstock strongly arched before dive
- does not raise flukes before dive

The Minke is the smallest and most abundant of the rorqual whales. It is slim, with a sharply pointed head, and shows relatively little of itself as it blows, surfaces and rolls through the water. Its smaller size compared with other rorqual whales provides the first indication of the species, and the way the tip of its snout breaks the surface first and at a slight angle is another identification feature. Named after an eighteenth-century Norwegian whaler, it is the only baleen whale still being regularly hunted by commercial whalers. Estimates of the population are therefore controversial, but they range from 500,000 to 1,000,000 individuals worldwide. Minkes are quite variable in appearance: animals in the northern hemisphere, for example, usually have a white band on their flippers, but this is absent from many southern hemisphere animals. They tend to take little notice of boats or people but, in some parts of the world, have learnt to recognize whalewatch vessels and often approach to within a few metres.

Distribution map

At a glance

Alternative names: Little Finner, Sharp-headed Finner, Pike Whale, Little Piked Whale, Pikehead, Lesser Finback, Lesser Rorqual.

Scientific name: *Balaenoptera acutorostrata*.

Adult size: 7–10m (23–33ft); 5–15 tonnes; females slightly larger than males.

Diet: shrimp-like krill and small schooling fish.

Behaviour: normally difficult to approach, but some individuals inquisitive; sometimes spyhops and breaches.

Distribution: virtually worldwide from the tropics to the edge of the polar ice, although most common in cooler waters; Atlantic population likely to be separate species.

Dive sequence

Spyhopping

Breaching

Sei Whale

The least known of all the rorqual whales, the Sei Whale usually lives far from shore, tends to be elusive and does not seem to gather in the same specific areas year after year. No commercial whalewatch operation is dedicated to watching this species, although it is sometimes encountered during tours specializing in Humpbacks, Minkes and other large whales. Named after a Norwegian word for the fish we call Pollack, for many years it was confused with the superficially similar Bryde's Whale (which has three ridges on the top of its head instead of the single ridge characteristic of the Sei Whale). As with all six members of the rorqual family, Sei Whales in the southern hemisphere tend to be larger than those in the northern hemisphere. They were heavily exploited by commercial whalers, especially during the 1960s and early 1970s, and the population has been severely depleted.

Identification

- relatively low, narrow blow
- single longitudinal ridge on head
- both sides of head evenly dark
- tall, sickle-shaped dorsal fin
- blowholes and fin visible simultaneously
- regular dive sequence
- does not arch tailstock before dive
- does not raise flukes before dive

At a glance

Alternative names: Sardine Whale, Pollack Whale, Coalfish Whale, Japan Finner, Rudolphi's Rorqual.

Scientific name: *Balaenoptera borealis*.

Adult size: 12–16m (39½–52½ft); 20–30 tonnes; females slightly larger than males.

Diet: small crustaceans, such as shrimp-like krill and copepods, and schooling fish.

Behaviour: regular dive sequence and remains near the surface between blows; seldom breaches.

Distribution: summer feeding grounds in deep, temperate waters worldwide, but believed to migrate into warmer, lower latitudes for winter; rarely seen close to shore, except in deep water around islands.

Distribution map

Dive sequence

Breaching

Bryde's Whale

Identification

- fairly tall, narrow blow
- three parallel ridges on head
- prominent, falcate dorsal fin
- skin may be mottled or scarred
- irregular dive sequence
- tailstock arched before dive
- does not raise flukes before dive

Unlike most other large whales, Bryde's Whales do not migrate long distances between separate feeding and breeding grounds each spring and summer, preferring to stay in warm tropical and sub-tropical waters with temperatures higher than 20°C (68°F) year-round. They make only short migrations – or none at all – and never visit cold waters. Named after a Norwegian consul, Johan Bryde, who helped to build the first whaling factory in Durban, South Africa, in 1909, Bryde's Whale is unique in having three parallel, longitudinal ridges on its head; all other members of the rorqual family have just one, running from the blowholes to the tip of the snout. There may be separate inshore and offshore populations in different parts of the world, differing slightly in appearance and behaviour, and there seem to be variations in the characteristics of Bryde's Whales from one geographical locality to another. Two separate species have now been proposed: Pygmy and Common.

Distribution map

At a glance

Alternative names: Tropical Whale.
Scientific name: *Balaenoptera edeni*.
Adult size: 11.5–14.5m (37½–47½ft); 12–20 tonnes; females slightly larger than males.
Diet: mainly schooling fish, but also squid and crustaceans.
Behaviour: sometimes inquisitive and may approach whalewatch boats; breaching common in some areas (the whale often leaves the water almost vertically and may arch its back in mid-air).
Distribution: warm waters worldwide, mainly between 30°N and 30°S, but does occur outside this broad range.

Dive sequence

Breaching

Fin Whale

Fin Whales are unusual in having asymmetrical pigmentation on their heads. The lower 'lip', mouth cavity and some of the baleen plates are white on the right side, but they are uniformly grey on the left side. No-one knows the reason for such intriguing coloration, but it is likely to be an adaptation for feeding, perhaps to confuse small prey. The second-largest living animal after the Blue Whale, the Fin Whale is a sleek and fast swimmer, capable of reaching speeds of over 30km/h (19mph). It tends to be more social than other members of the rorqual family, and is often seen in small groups, typically of three to seven individuals. Once one of the most abundant of the large whales, the whaling industry hunted it ruthlessly and its population has been severely depleted. Native whaling still occurs off Greenland.

Identification

- very tall, narrow blow
- exceptionally large size
- backward-sloping dorsal fin
- single longitudinal ridge on head
- asymmetrical coloration on head
- greyish white chevron
- rarely raises flukes before dive

At a glance

Alternative names: Finback, Finner, Herring Whale, Common Rorqual, Razorback.

Scientific name: *Balaenoptera physalus*.

Adult size: 18–26m (59–85½ft); 30–80 tonnes; females slightly larger than males.

Diet: variety of schooling fish, krill and other crustaceans and, to lesser extent, squid.

Behaviour: usually indifferent to boats, neither avoiding them nor approaching them; rarely breaches.

Distribution: deep water in tropical, temperate and polar regions worldwide, but most common in cooler waters; some populations seem to be resident year-round, but others may migrate between warm waters in winter and cooler waters in summer (movements less predictable than in some other large whales).

Distribution map

Dive sequence

Breaching

Blue Whale

Identification

- very tall, narrow blow
- exceptionally large size
- blue-grey body colour with mottling
- tiny, stubby dorsal fin set far back
- broad, flattened, U-shaped head
- huge blowhole splashguard
- extremely thick tailstock
- may raise flukes on diving

The largest living animal on Earth, the Blue Whale is almost as long as a Boeing 737 and weighs nearly as much as 2,000 people. A length of more than 33m (110ft) and a weight of nearly 200 tonnes have been recorded, although these are exceptional. The Blue Whale needs so much food that, in terms of weight, it could eat a fully-grown African Elephant every day. Sadly, its sheer size made it one of the most sought-after whales during the heyday of modern whaling and hundreds of thousands of them were killed worldwide. As a result, some stocks may never recover. There are believed to be three different sub-species: the largest lives in the Antarctic; a slightly smaller one lives in the northern hemisphere; and an even smaller one, known as the Pygmy Blue Whale, lives mainly in tropical waters of the southern hemisphere. The characteristic mottling of the back and sides picks up reflected blues of the sea and sky, adding an extra dimension to the basic blue-grey body colour.

Distribution map

At a glance

Alternative names: Sulphur-bottom, Sibbald's Rorqual, Great Northern Rorqual.

Scientific name: *Balaenoptera musculus*.

Adult size: 21–27m (69–88½ft); 100–120 tonnes; females larger than males.

Diet: shrimp-like krill, with some squid, amphipods, copepods and red crabs.

Behaviour: some individuals are approachable; breaching is known in juveniles, but rarely observed in adults.

Distribution: worldwide, although distribution is very patchy; some populations migrate between low-latitude winter breeding grounds and high-latitude summer feeding grounds, but others appear to be resident year-round; very small numbers in Europe.

Dive sequence

Humpback Whale

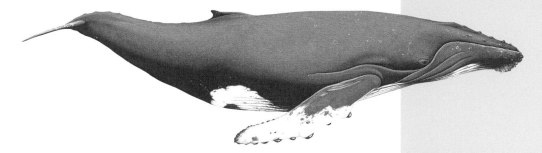

Herman Melville, the author of *Moby Dick*, described the Humpback Whale as 'the most gamesome and lighthearted of all the whales, making more gay foam and whitewater generally than any of them.' Its spectacular breaching, as well as lobtailing, flipper-slapping and spyhopping, make it particularly popular with whalewatchers. It is also among the most studied of all the world's cetaceans. Biologists use the distinctive black and white markings on the underside of the tail flukes to tell one individual from another, and they have prepared huge catalogues containing photographs of thousands of individually-identified Humpbacks around the world. Humpback Whales are easy to identify at close range, when their enormous flippers and the knobs or tubercles (golf ball-sized hair follicles on the rostrum and lower jaw) are visible. However, they can be confused with other large whales at a distance.

Identification

- black or dark grey upperside
- varying amounts of white on underside
- stocky body
- low, stubby dorsal fin sits on hump
- long pectoral fins (up to one-third of body length)
- knobs on rostrum and lower jaw
- black and white markings on underside of flukes
- flukes raised before deep dive
- wide, bushy blow
- alone, small groups or up to 15 together

At a glance

Alternative names: Hump-backed Whale (rare).
Scientific name: *Megaptera novaeangliae.*
Maximum adult size: 15m (49½ft) male, 16.5m (54ft) female (reputed record of 19m (62½ft) questioned by many experts); 35–45 tonnes.
Diet: crustaceans and small schooling fishes such as Herring, Capelin and Sand Lance.
Behaviour: inquisitive and approachable with lots of surface activity.
Distribution: wide-ranging worldwide, but with distinct seasonal changes; winter in low latitude breeding grounds, summer in high latitude feeding grounds.

Distribution map

Dive sequence   Fluking   Breaching   Flipper-   Spyhopping
                                                                        slapping

Sperm Whale

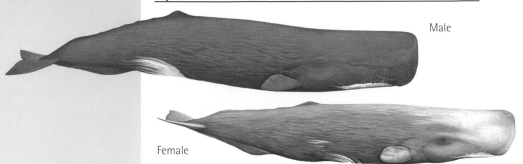

Male

Female

Identification

- bushy blow projected forward and to left
- dark body with wrinkled skin
- huge, squarish head
- low hump in place of fin
- 'knuckles' from hump to flukes
- single, slit-like blowhole
- often lies motionless at surface
- broad flukes raised on diving

Easily recognized by their huge, squarish heads, wrinkly prune-like skin and uniquely angled, bushy blows, Sperm Whales behave more like submarines than air-breathing mammals. Capable of diving deeper and for longer than any other mammal, they have been tracked at depths of 2,000m (6,560ft) and there is circumstantial evidence to suggest that they may be able to dive to 3,000m (9,840ft) or even deeper. These amazing forays into the cold, dark ocean depths can last for two hours or more. The two sexes normally live together during the breeding season only. At other times, there are two main groupings: 'bachelor schools', containing non-breeding males, and 'nursery schools', containing females with calves of both sexes. Older males sometimes live alone. The Sperm Whale was once the mainstay of the whaling industry, and huge numbers were killed over several centuries. It somehow survived, however, and is now probably the most abundant of all the great whales.

Distribution map

At a glance

Alternative names: Cachalot, Great Sperm Whale.
Scientific name: *Physeter macrocephalus.*
Adult size: 11–18m (36–59ft); 20–50 tonnes; mature males can be one and a half times the length of mature females.
Diet: mainly deep-water squid of all sizes, although octopuses and large fish also taken.
Behaviour: often unconcerned about whalewatch boats; frequently breaches (especially juveniles).
Distribution: patchy distribution in deep water worldwide, from tropics to sub-polar waters; normally, only large males venture to extreme north and south of range (females rarely venture beyond 45°N or 42°S).

Dive sequence

Fluking

Breaching

Spyhopping

Northern Bottlenose Whale

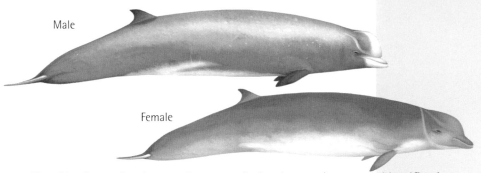

Male

Female

The Northern Bottlenose is one of the better known beaked whales and is certainly one of the most inquisitive. Both sexes regularly approach ships, and this behaviour has made them more likely to be studied and, more recently, whalewatched than any other member of the family. They also have been hunted more than any other beaked whale and tens of thousands have been killed since the late nineteenth century; scientists disagree about the extent to which the species has been depleted, and its current status. They are deep divers, known to venture to depths of 1,000m (3,280ft) or more, and use their superb sonar system to pursue deep-water squid. Older males, in particular, are very distinctive with their large bulbous foreheads and prominent tube-like beaks. Research in the western North Atlantic suggests that they are resident year-round, but at least some populations in Europe appear to be migratory.

Identification

- visible bushy blow
- dark grey to brown colour
- robust, cylindrical body
- dorsal fin two-thirds of way down back
- bulbous forehead
- prominent tube-like beak
- no notch in flukes

At a glance

Alternative names: North Atlantic Bottlenosed Whale, Flathead, Bottlehead, Steephead.
Scientific name: *Hyperoodon ampullatus.*
Adult size: 7–9m (23–29½ft); 5.8–7.5 tonnes; males larger than females.
Diet: mainly squid, but also some shoaling fish and invertebrates.
Behaviour: often quite curious and will approach stationary boats; breaching not uncommon.
Distribution: deep, cold temperate waters in the northern N. Atlantic, mainly along or beyond the edge of the continental shelf and over submarine canyons.

Distribution map

Dive sequence

Spyhopping

Dwarf Sperm Whale

Identification

- low, inconspicuous blow
- small, robust body
- squarish head
- prominent falcate dorsal fin
- false gill behind each eye
- may float motionless at surface
- simply drops below surface

The Dwarf Sperm Whale is the smallest of the whales and is even smaller than some dolphins. Its square head and slow, deliberate movements distinguish it from Bottlenose and other dolphins, but it is often confused with the Pygmy Sperm Whale. The two species were not officially separated until 1966, although they can be distinguished at close range by their dorsal fins (the fin of the Dwarf Sperm Whale is much larger and more erect – somewhat like a Bottlenose Dolphin's dorsal fin). The two species share several intriguing characteristics: their underslung lower jaws and the creamy-white arcs behind each eye (that resemble gill-slits) make them look superficially like sharks; when resting between dives, they tend to float motionless at the surface with part of the head and back exposed and the tail hanging limply in the water; and when frightened, they sometimes evacuate a reddish-brown faecal material, which may function as a decoy, like squid ink.

Distribution map

At a glance

Alternative name: Owen's Pygmy Sperm Whale.
Scientific name: *Kogia simus*.
Adult size: 2.1–2.7m (7–9ft); 135–275kg (300–605lb).
Diet: deep-water squid and octopus, but will also take variety of fish, cuttlefish and crustaceans.
Behaviour: usually shy and rarely approaches boats; rises to the surface slowly and deliberately and, unlike most other small whales, simply drops out of sight.
Distribution: worldwide in tropical, sub-tropical and warm temperate waters, although never officially recorded across vast areas within this assumed range; predominantly occurs in deep water.

Dive sequence

Pygmy Sperm Whale

With a preference for deep water far from shore, and rather inconspicuous habits, the Pygmy Sperm Whale can be a difficult animal to find. It is often confused with the Dwarf Sperm Whale, which was not recognized as a separate species until 1966, although it can be identified at close range by its much smaller dorsal fin. Stranded animals have also been mistaken for sharks, because of their underslung lower jaws and some unusual creamy-white markings on either side of the head that resemble gill-slits. The Pygmy Sperm Whale is a deep diver. When resting between dives, it tends to float motionless at the surface with part of the head and back exposed and the tail hanging limply in the water. It is easily frightened and may evacuate a reddish-brown faecal material, leaving behind a dense cloud in the water while it dives to safety; this may function as a decoy, like squid ink.

Identification

- low, inconspicuous blow
- small, robust body
- squarish head
- tiny, hooked, falcate dorsal fin
- false gill behind each eye
- may float motionless at surface
- simply drops below surface

At a glance

Alternative names: Lesser Sperm Whale, Short-headed Sperm Whale, Lesser Cachalot.
Scientific name: *Kogia breviceps.*
Adult size: 2.7–3.4m (9–11½ft); 315–400kg (695–880lb); males slightly larger than females.
Diet: deep-water squid and octopus, but will also take variety of fish, cuttlefish and crustaceans.
Behaviour: usually shy and rarely approaches boats; rises to the surface slowly and deliberately and, unlike most other small whales, simply drops out of sight.
Distribution: worldwide in tropical, sub-tropical and warm temperate waters, although never officially recorded across vast areas within this assumed range.

Distribution map

Dive sequence

Breaching

Beluga

Identification

- robust body shape
- white, creamy-white or yellowish body colour
- relatively small head with rounded melon
- broad mouthline
- distinct and flexible neck
- dorsal ridge instead of dorsal fin

Ancient mariners used to call the Beluga the 'sea canary' because of its great repertoire of trills, moos, clicks, squeaks and twitters. One of the most vocal of the toothed whales, it can even be heard from above the surface or through the hull of a boat. It can also make a variety of facial expressions by altering the shape of its forehead and lips. Not all Belugas are white: the body colour changes with age, from dark slate-grey at birth to pure white when the animal is sexually mature at five or ten years old. Well adapted to living close to shore, Belugas are able to manoeuvre in very shallow water and, if stranded, can often survive until the next tide refloats them (unless they are found by a Polar Bear – many Belugas have scars caused by unsuccessful bear attacks). Belugas have been hunted for centuries, but greater concerns come from the effects of oil and gas activities and chemical pollution.

At a glance

Alternative names: Belukha, Sea Canary, White Whale.
Scientific name: *Delphinapterus leucas.*
Adult size: 3–5m (9½–16½ft); 0.4–1.5 tonnes; males larger than females.
Diet: wide variety of fish, as well as crustaceans, squid, octopuses and molluscs.
Behaviour: fairly easy to approach; frequently spyhops and shows curiosity towards boats.
Distribution: mainly in seasonally ice-covered waters of sub-Arctic and Arctic and in shallow coastal waters; will enter estuaries and even travel hundreds of kilometres up rivers.

Distribution map

Dive sequence

Fluking   Spyhopping   Side rolling

Narwhal

Male

Female

With its long, spiralling tusk, looking like a gnarled and twisted walking stick, the male Narwhal is unique and unlikely to be confused with any other cetacean. Until early in the seventeenth century the tusk was believed to be the horn of the legendary unicorn, but it is actually a modified tooth. Its role baffled scientists for years and among many theories were that it was used for spearfishing, grubbing for food and drilling through ice. In fact, it is probably used in fights over females and as a visual display of strength. A small number of males have two tusks and, rarely, females grow them as well. Narwhals change colour as they grow older – blotchy grey or brownish grey when they are first born, then uniformly purplish black, then black or dark brown smudges on a grey background, and finally very old animals can be almost entirely white. They have been hunted for centuries by indigenous peoples and are still hunted today in northern Canada and Greenland.

Identification

- long tusk of male
- mottled back and sides
- slight hump instead of dorsal fin
- bulbous forehead
- short, upcurled flippers
- 'backward-facing' flukes
- inhabits very high latitudes

At a glance

Alternative name: Narwhale.

Scientific name: *Monodon monoceros.*

Adult size: 3.8–5m (12½–16½ft) not including male's tusk (up to 3m (10ft)); 0.8–1.6 tonnes; males slightly larger than females.

Diet: wide variety of fish, squid and crustaceans.

Behaviour: males joust with crossed tusks; spyhopping, lobtailing and flipper-slapping fairly common, but breaching rare.

Distribution: truly Arctic species, living farther north than almost any other cetacean; mostly above Arctic Circle, and right to edge of ice-cap, but also in adjacent bays and straits.

Distribution map

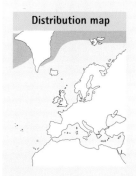

Dive sequence

Jousting males   Spyhopping

Sowerby's Beaked Whale

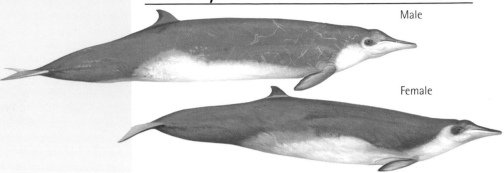

Male

Female

Identification

- long, slender body
- bluish-grey or slate-grey upperside, lighter underside
- limited scarring and blotching
- relatively small head
- long, slender beak with straight mouthline
- two teeth mid-way along lower jaw (male only)
- bulge in front of blowhole
- small, curved dorsal fin
- dorsal fin far behind centre

Sowerby's Beaked Whale was the first of the beaked whales to be named. A lone male was found stranded in the Moray Firth, Scotland, in 1800, and its skull was collected. Four years later the species was described by the English watercolour artist James Sowerby. It has one of the most northerly distributions of all the beaked whales, although parts of its range do overlap with other *Mesoplodon* species and identification can therefore be extremely difficult. The position of the two teeth in the male is distinctive, lying mid-way along the mouthline, and these are visible even when the mouth is closed – but only at close range. Although Sowerby's was the first beaked whale to be named, it remains one of the more elusive members of the family and is still poorly known. It is sometimes called the North Sea Beaked Whale because, although its range extends across the North Atlantic, it is most likely to be found in the northern North Sea.

Distribution map

At a glance

Alternative name: North Sea Beaked Whale.
Scientific name: *Mesoplodon bidens.*
Adult size: 4–5.5m (13$\frac{1}{2}$–18ft); 1–1.3 tonnes; males slightly larger than females.
Diet: deep-water squid and small fish.
Behaviour: generally unobtrusive and apparently does not approach boats; some reports suggest head is brought out of the water at a steep angle when surfacing.
Distribution: cold temperate and sub-Arctic waters of the northern N. Atlantic, with most records from the east; mainly deep waters offshore.

Dive sequence (artist's impression, based on reported sightings)

Blainville's Beaked Whale

Male

Female

Blainville's Beaked Whale has a wide distribution, but sightings are still relatively rare and it remains poorly known. The male is one of the oddest-looking of all cetaceans. It has a pair of massive teeth that grow from substantial bulges in its lower jaw, like a couple of horns, and these may be so encrusted with barnacles that the animal appears to have two dark-coloured pompons on top of its head. The females have less prominently arched lower jaws and, as with most other beaked whales, their teeth do not erupt. This species was named in 1817 by Henri de Blainville, who described it from a small piece of jaw, which was the densest bone structure he had ever seen. It is often known as the Dense-beaked Whale and is reputed to have the densest bones in the animal kingdom – denser even than elephant ivory.

At a glance

Alternative names: Tropical Beaked Whale, Dense-beaked Whale, Atlantic Beaked Whale.

Scientific name: *Mesoplodon densirostris*.

Adult size: 4.5–6m (14½–19½ft); c.1 tonne; males larger than females.

Diet: mainly deep-water squid, but possibly some fish.

Behaviour: generally unobtrusive and difficult to find; limited reports suggest that, on surfacing, beak appears first at a sharp angle.

Distribution: warm temperate to tropical waters, predominantly around the US Atlantic coast but with patchy records from Europe and many other parts of the world; mainly deep waters offshore.

Identification

- robust, spindle-shaped body
- dark upperside, paler underside
- scars and blotches all over body (especially males)
- strongly arched lower jaw
- huge, forward-tilting, horn-like teeth (male only)
- flattened forehead with depression between raised teeth
- thick, moderately long beak
- prominent curved or triangular dorsal fin
- dorsal fin far behind centre

Distribution map

Dive sequence

Gervais' Beaked Whale

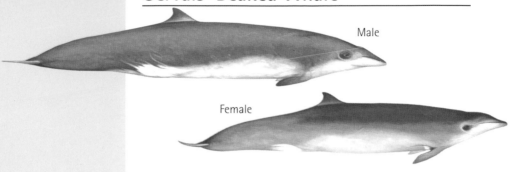

Male

Female

Identification

- spindle-shaped body
- dark grey or marine blue upperside, pale grey underside
- extensive body scarring
- white blotches on underside
- slightly bulging forehead
- two tiny teeth set back from tip of jaw (male only)
- indentation at blowhole
- small, shark-like dorsal fin
- dorsal fin far behind centre

Distribution map

The first recorded specimen of Gervais' Beaked Whale was found floating in the English Channel in the 1840s (hence its scientific name, *europaeus*, and the alternative name European Beaked Whale) but there have been very few records in Europe since. It is the most commonly stranded *Mesoplodon* species along the Atlantic coast of the USA and appears to be widely distributed in the Caribbean, but records from other parts of the world are so widely dispersed that it is difficult to tell if there is another important centre of distribution. A stranded male can be identified by the single pair of teeth, which are located about one-third of the way from the tip of the beak to the corner of the mouth, although these would be difficult to spot at sea; the teeth do not erupt in females.

At a glance

Alternative names: Gulf Stream Beaked Whale, European Beaked Whale, Antillean Beaked Whale.
Scientific name: *Mesoplodon europaeus*.
Adult size: 4.5–5.2m (14½–17ft); c.1–2 tonnes.
Diet: deep-water squid.
Behaviour: lack of sightings in relatively well-studied areas within range suggests it is likely to be inconspicuous.
Distribution: deep sub-tropical and warm temperate waters in western N. Atlantic, with scattered records in the eastern N. Atlantic and S. Atlantic.

Dive sequence (artist's impression, based on reported sightings)

True's Beaked Whale

Male

Female

In 1913 the American biologist Frederick True named this species *mirus*, meaning 'wonderful'. True's Beaked Whale has a special place in the hearts of European whalewatchers, because occasional strandings have proved that it occurs in the region but, until recently, it had not been positively identified at sea. However, it has now been observed in the Azores and, on 9 July 2001, in the southern Bay of Biscay, surveyors from Organisation Cetacea witnessed an adult male breach 24 times in a row alongside their ship – and managed to get photographic evidence. As with most other *Mesoplodon* species, females and juveniles are probably unidentifiable, unless stranded specimens can be examined closely. There may be two forms of True's Beaked Whale – one in the North Atlantic and the other in the southern hemisphere – with slight cranial and pigmentation differences between them.

At a glance
Alternative name: Wonderful Beaked Whale.
Scientific name: *Mesoplodon mirus.*
Adult size: 4.9–5.3m (16–17½ft); c.1–1.5 tonnes.
Diet: deep-water squid.
Behaviour: with just a handful of possible sightings, virtually nothing is known about behaviour at sea; reported to breach.
Distribution: mostly known from temperate waters in the western N. Atlantic, but also recorded in the eastern N. Atlantic, and in South Africa and Australasia.

Identification

- predominantly grey or bluish-grey colour
- may be paler towards tail and on underside
- body scratched and scarred
- dark patch around each eye
- relatively short beak
- two small teeth at tip of jaw (male only)
- slightly bulging forehead
- slight indentation at blowhole
- small, curved dorsal fin far behind centre

Distribution map

Dive sequence (artist's impression, based on reported sightings)

Cuvier's Beaked Whale

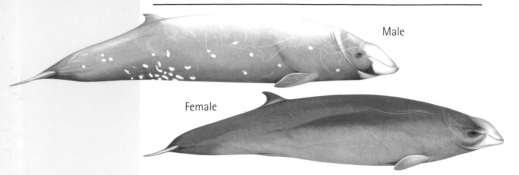

Male

Female

Identification

- robust body
- body colour highly variable
- long and circular scars
- small head, often pale
- forehead slopes gently to slight beak
- indentation behind blowhole
- small teeth at tip of jaw (male only)
- dolphin-like dorsal fin far behind centre

When Cuvier's Beaked Whale was first described in 1823, by the French anatomist Georges Cuvier, it was believed to be extinct. However, now it is recognized as one of the most widespread and abundant of the beaked whales and is probably one of the three most watched beaked whales in the world (along with the Northern Bottlenose Whale and Baird's Beaked Whale). Older males sometimes travel alone, but small groups are more typical and as many as 25 have been seen together. The body colour varies from individual to individual, but is usually brown with a large creamy white area over the head and front portion of the body; males tend to have a larger pale area than females and are often heavily scarred from the teeth of other males. The shape of the head and beak is sometimes described as resembling a goose's beak, which is why its alternative name is the Goose-beaked Whale.

At a glance

Alternative names: Goose-beaked Whale, Goosebeak Whale, Cuvier's Whale.
Scientific name: *Ziphius cavirostris.*
Adult size: 5.5–7m (18–23ft); 2–3 tonnes.
Diet: deep-water squid and fish.
Behaviour: can be indifferent to boats or inquisitive and approachable; known to breach.
Distribution: widely distributed in cold temperate to tropical waters worldwide, especially around oceanic islands and in enclosed seas; mainly deep waters offshore and often associated with submarine canyons and escarpments on the continental shelf edge.

Distribution map

Dive sequence

Breaching

Killer Whale

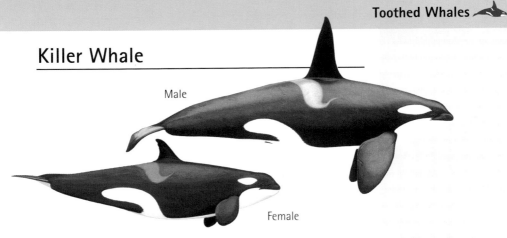

Male

Female

Centuries ago, when Basque whalers saw Killer Whales feeding on the carcasses of dead whales, they called them 'whale killers'. Some say that a translator made a mistake and reversed the words, forming the name we use today. Killer Whales have had a bad reputation ever since, and yet they do not deserve their killer name any more than other top predators such as Lions, Tigers and Polar Bears. In fact, they are unusual because they do not hunt people and, from a human perspective, are far less dangerous than many other animals of a similar size. They live in close-knit family groups, known as pods, which bridge the generation gap and are usually so stable only death or capture can break them apart. The largest members of the dolphin family, they are easily recognized by their predominantly jet-black and brilliant-white markings. The enormous dorsal fin of the male – which can be roughly the same height as a human – is also distinctive.

Identification

- distinctive black and white coloration
- huge triangular dorsal fin of male
- robust, heavy body
- blunt snout with indistinct beak
- large, paddle-shaped pectoral fins
- usually in mixed family groups

At a glance
Alternative names: Orca.
Scientific name: *Orcinus orca.*
Adult size: average 6.7m (22ft) in male, 5.8m (19ft) in female.
Diet: includes squid, fishes, seabirds, sea turtles, seals and sealions, sea otters, cetaceans; family pods tend to specialize.
Behaviour: inquisitive and approachable with lots of surface activity.
Distribution: wide-ranging but patchy worldwide; most common in cooler, high latitude waters but also found in warmer, low latitude waters.

Distribution map

Dive sequence

Lobtailing

Breaching

Spy-hopping

Hunting

Cat and mouse play

Short-finned Pilot Whale

Female

Male

Identification

- strong, low bushy blow
- jet black or dark grey/brown colour
- stocky but elongated body
- rounded, bulbous forehead
- large, broad-based dorsal fin
- dorsal fin set far forward on body

Short-finned Pilot Whales are almost impossible to tell apart from Long-finned Pilot Whales at sea but, fortunately, there is relatively little overlap in their range. The main differences are in the length of the flippers, the shape of the skull and the number of teeth. They live in close-knit family groups, typically containing 10 to 30 but sometimes as many as 60 individuals, and are such social animals that they are almost never seen alone. When travelling, pods often swim abreast in long 'chorus lines', and they are often accompanied by Bottlenose Dolphins and other small cetaceans. In some parts of the world, they may also be shadowed by Oceanic Whitetips and other pelagic sharks, although the nature of this association is poorly understood. Studies in the Canary Islands suggest that they prefer water depths of about 1,000m (3,300ft), where they make deep dives mainly to feed on squid. They are long-lived animals, with females known to survive up to 65 years.

Distribution map

At a glance
Alternative names: Shortfin Pilot Whale, Pacific Pilot Whale, Blackfish, Pothead Whale.
Scientific name: *Globicephala macrorhynchus*.
Adult size: 5–6.5m (16½–21½ft); 1–4 tonnes; males slightly larger than females.
Diet: mainly squid, although will take small and medium-sized fish when opportunities arise.
Behaviour: frequently lobtails and spyhops; pods sometimes rest motionless at the surface, allowing boats to approach closely.
Distribution: warmer waters worldwide, mainly in deep water; little, but probably some, overlap in range with Long-finned Pilot Whale.

Dive sequence

Spyhopping

Lobtailing

Lying on one side

Long-finned Pilot Whale

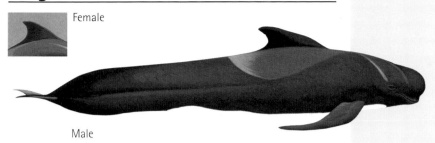

Female

Male

The Long-finned Pilot Whale is a family animal, travelling in groups of 10 to 50, and sometimes up to 100 or more. There are reports of thousands seen together in great superpods and they are often found in the company of other small cetaceans. Well-known for stranding on beaches, they are possibly subject to more mass strandings than any other cetacean. They have been hunted around the Faroe Islands, halfway between Scotland and Iceland in the north-east Atlantic, for centuries. During noisy drives that frequently take many hours to complete, entire pods are herded into sandy bays by men in small boats, then dragged ashore with steel hooks and killed with long knives. Long-finned and Short-finned Pilot Whales are almost impossible to tell apart at sea but, fortunately, there is relatively little overlap in their range. The main differences are in the length of the flippers, the shape of the skull and the number of teeth.

At a glance
Alternative names: Caaing Whale, Longfin Pilot Whale, Atlantic Pilot Whale, Blackfish, Pothead Whale.
Scientific name: *Globicephala melas*.
Adult size: 3.8–7.5m (12½ –24½ft); 1.8–3.5 tonnes; males larger than females.
Diet: mainly squid, although will take small and medium-sized fish when opportunities arise.
Behaviour: frequently lobtails and spyhops; pods sometimes rest motionless at surface, allowing boats to approach closely.
Distribution: two distinct populations in deep cold temperate to sub-polar waters: one in the southern hemisphere and the other in the N. Atlantic; little, but probably some, overlap in range with Short-finned Pilot Whale.

Identification
- strong, low bushy blow
- jet black or dark grey/brown colour
- stocky but elongated body
- rounded, bulbous forehead
- large, broad-based dorsal fin
- dorsal fin set far forward on body

Distribution map

Dive sequence

Spyhopping   Lobtailing

Lying on one side

45

Pygmy Killer Whale

Identification

- robust, dark-coloured body
- darker dorsal cape
- rounded head with no beak
- white 'lips'
- some individuals have white chin
- large white patch on belly
- prominent, falcate fin

Although no bigger than a dolphin itself, the Pygmy Killer Whale has been reported to herd and attack dolphins in the South Atlantic and the tropical Pacific and, in captivity, has been quite aggressive towards people. It is very similar to the Melon-headed Whale, with which it shares almost the same habitat and range, and the two species can be difficult to distinguish at sea. The Pygmy Killer has a more rounded head and flipper tips, a slightly broader body and a darker cape, although these features can only be seen sufficiently clearly when the animals are bow-riding; skull comparisons of stranded specimens also show fewer teeth. Some individuals have a distinctive white 'chin', while others merely have white 'lips', but, again, these are only visible at close range. Although Pygmy Killer Whales are widely distributed, surprisingly little is known about them and they are rarely encountered on whalewatching trips. A glimpse of a pod, usually swimming abreast in a perfectly coordinated 'chorus line', is an unusual and welcome sight.

Distribution map

At a glance
Alternative names: Slender Blackfish, Slender Pilot Whale.
Scientific name: *Feresa attenuata*.
Adult size: 2.1–2.6m (7–8½ ft); c.110–170kg (240–375lb).
Diet: mostly fish and squid, although known to attack other dolphins.
Behaviour: lively swimmer, but aerial behaviour is quite rare; may bow-ride and wake-ride but some populations tend to avoid boats.
Distribution: deep, mainly offshore waters in sub-tropics and tropics; rarely occurs close to shore, except around oceanic islands.

Dive sequence

Spyhopping

Fast swimming

Lobtailing

Melon-headed Whale

Despite its wide distribution, the Melon-headed Whale is rarely encountered at sea and is relatively poorly known. It was originally thought to be a dolphin, but when scientists studied a large herd that had been caught and killed off Japan, in 1965, it was found to be so different that it deserved its own genus. Named for its pointed, melon-shaped head, it often creates a lot of spray as it surfaces and frequently changes direction underwater, so can be quite a challenging animal to observe in detail. It is very similar to the Pygmy Killer Whale, with which it shares almost the same habitat and range, and the two species can be difficult to distinguish at sea. Generally speaking, if a large number of animals are seen together (more than 50) they are more likely to be Melon-headed Whales. They normally travel in tightly-packed herds of 100 to 500, although as many as 2,000 have been seen together. In some parts of the world, they show little fear of boats or people.

Identification

- torpedo-shaped, dark-coloured body
- slim, pointed head with no beak
- dark 'mask' on face
- white, light grey or pinkish 'lips'
- tall, falcate dorsal fin
- long, sharply pointed flippers

At a glance

Alternative names: Little Killer Whale, Electra Dolphin, Melonhead Whale, Many-toothed Blackfish.
Scientific name: *Peponocephala electra*.
Adult size: 2.1–2.7m (7–9ft); c.160kg (355lb).
Diet: squid and various small fish.
Behaviour: may bow-ride and spyhop and will approach boats in some parts of the world.
Distribution: sub-tropical and tropical waters worldwide; usually encountered offshore, or around oceanic islands, and seldom ventures close to land.

Distribution map

Dive sequence

Fast swimming

False Killer Whale

Identification

- uniformly dark colour
- long, slim body
- slender head tapering to rounded beak
- head all black with no white 'lips'
- prominent dorsal fin
- unique 'elbow' on flippers
- flukes small in relation to body size

Distribution map

The False Killer Whale can look rather menacing, with its black torpedo-shaped body and rows of sharp teeth, and, indeed, it is known to attack groups of small cetaceans. Yet, in many ways, it behaves more like a dolphin – leaping high into the air, making rapid turns underwater, and even riding the bow-waves of passing boats and ships. Most pods are relatively small (10 to 50 animals), although several hundred have been seen travelling together, and they include both sexes and all ages. False Killer Whales are often involved in mass strandings; the largest recorded was of more than 800 animals. Their size distinguishes them from Pygmy Killer Whales and Melon-headed Whales, and they are slimmer and have a more dolphin-like dorsal fin than the Short-finned Pilot Whale.

At a glance

Alternative names: False Pilot Whale, Pseudorca.
Scientific name: *Pseudorca crassidens.*
Adult size: 4.3–6m (14–19½ft); 1.1–2.2 tonnes; males slightly larger than females.
Diet: mainly fish and squid, although also known to attack dolphins and there is one record of them attacking and killing a Humpback Whale calf near Hawaii.
Behaviour: will approach boats to investigate, bow-ride or wake-ride; highly acrobatic and often breaches, causing a huge splash for a whale of its size.
Distribution: widely distributed in tropical, sub-tropical and sometimes warm temperate waters worldwide; prefers deep water and normally encountered offshore.

Dive sequence

High breaching

Rough-toothed Dolphin

A strange-looking animal, with a long narrow beak that blends into the forehead without a crease, the Rough-toothed Dolphin has a slightly reptilian or primitive appearance. It has been described as the 'ugly duckling' of the dolphin world, but has its own unique beauty. It is named for a series of fine, vertical wrinkles on the enamel cap of each tooth, although these are impossible to see in the wild. In contrast, the light-coloured blotches on some individuals are quite distinctive at sea and may be caused by the bites of Cookiecutter Sharks. Rough-toothed Dolphins are gregarious, normally travelling in groups of 10–20, and sometimes up to 50 or more at a time. Although they have a reputation as deep divers and fast swimmers, frequently porpoising with low, arc-shaped leaps, they have also been observed cruising around slowly at the surface.

At a glance
Alternative name: Slopehead.
Scientific name: *Steno bredanensis.*
Adult size: 2.1–2.6m (7–8½ft); 100–150kg (220–330lb).
Diet: fish, squid and octopus, and possibly molluscs.
Behaviour: occasionally bow-rides in front of fast-moving vessels.
Distribution: mainly deep oceanic waters in warm temperate, sub-tropical and tropical waters worldwide; rarely ranges north of 40°N or south of 35°S or where sea surface temperature is below 25°C (77°F).

Identification

- robust body in front of dorsal fin (slimmer behind)
- dark, narrow cape
- pinkish or yellowish-white blotches (variable)
- conical head
- long beak continuous with sloping forehead
- white or pinkish white 'lips'
- tall, falcate dorsal fin
- large flippers set far back

Distribution map

Dive sequence

Fast swimming

Short-beaked Common Dolphin

Identification

- slender, streamlined body (slightly chunkier than Long-beaked)
- dark cape with V-shape under dorsal fin
- criss-cross or hourglass pattern on sides
- tan or yellowish patches on sides
- white underside and lower sides
- predominantly dark flippers, flukes and dorsal fin
- more crisp colour pattern than Long-beaked

There are many variations of Common Dolphin and more than 20 different species have been proposed and rejected over the years. They all have the distinctive hourglass pattern of white, grey, yellow and black on their sides, forming a dark V-shape below the dorsal fin, which looks almost like a reflection of the fin itself, but they show many differences within this basic framework. In 1995, though, the Common Dolphin was officially separated into two distinct species, now known as the Short-beaked Common Dolphin and the Long-beaked Common Dolphin. As their new names suggest, the most distinctive features can be seen in their beaks: the Short-beaked has a relatively short, slightly stubbier beak with more open markings than the Long-beaked. The Short-beaked also has a brighter and more contrasting colour pattern on its body, and there are other physical, genetic and behavioural differences.

At a glance

Alternative names: Criss-cross Dolphin, Saddleback Dolphin, Cape Dolphin, White-bellied Porpoise, Common Porpoise.
Scientific name: *Delphinus delphis.*
Adult size: 1.7–2.4m (5½–8ft); 70–110kg (155–245lb); males slightly larger than females.
Diet: small schooling fish and squid.
Behaviour: fast swimmers and energetic, boisterous acrobats; frequently bow-ride (when their high-pitched vocalizations can often be heard above the surface).
Distribution: worldwide in warm temperate, sub-tropical and tropical waters; mainly offshore.

Distribution map

Dive sequence

Spyhopping

Fast swimming

Long-beaked Common Dolphin

The Long-beaked Common Dolphin is by far the rarer of the two species of Common Dolphin in Europe. Its normal distribution in the North Atlantic is limited to the coasts of West Africa and northern South America, and it is rarely seen outside warm temperate and tropical waters. Where there is overlap in range, the two species can be difficult to tell apart at sea, although the Long-beaked has a longer, less stubby beak and a more muted colour pattern.

At a glance

Alternative names: Criss-cross Dolphin, Saddleback Dolphin, Cape Dolphin, White-bellied Porpoise, Common Porpoise.
Scientific name: *Delphinus capensis*.
Adult size: 1.7–2.4m (5½–8ft); 70–110kg (155–245lb); males slightly larger than females.
Diet: small schooling fish and squid.
Behaviour: fast swimmers and energetic, boisterous acrobats; frequently bow-ride (when their high-pitched vocalizations can often be heard above the surface).
Distribution: worldwide in warm temperate, sub-tropical and tropical waters; mainly coastal, but also offshore.

Identification

- slender, streamlined body (less chunky than Short-beaked)
- dark cape with V-shape under dorsal fin
- criss-cross or hourglass pattern on sides
- tan or yellowish patches on sides
- white underside and lower sides
- predominantly dark flippers, flukes and dorsal fin
- more muted colour pattern than Short-beaked

Distribution map

Dive sequence   Spyhopping   Fast swimming

Atlantic Spotted Dolphin

Identification

- fairly robust head and body
- most adults heavily spotted
- dark, purplish-grey cape, light grey underside
- long, chunky, white-tipped beak
- tall, falcate dorsal fin
- variable appearance within herd

No two Atlantic Spotted Dolphins look alike because, even in adults, the spotting varies greatly from one individual to another and from region to region. Young animals have no spots (they begin to appear as they grow older) and some elderly animals have so many large spots that the normal background colour of their bodies is barely visible. Spotting usually decreases from west to east across the Atlantic and with distance from the mainland. It can make identification at sea quite challenging in some areas, especially since it is not unusual for Bottlenose Dolphins to have a moderate number of spots as well. Two main forms of Atlantic Spotted Dolphin are recognized: inshore and offshore; the inshore form is larger, more robust and more heavily spotted. One particular population has been studied in great detail for more than 20 years – on Little Bahama Bank, north of The Bahamas – but the species is poorly known elsewhere in the Atlantic.

Distribution map

At a glance

Alternative names: Spotted Porpoise, Long-snouted Dolphin, Spotter, Bridled Dolphin, Gulf Stream Spotted Dolphin.
Scientific name: *Stenella frontalis*.
Adult size: 1.7–2.3m (5½–7½ft); 100–140kg (220–310lb).
Diet: fish and squid.
Behaviour: fast and energetic swimmer and very active at the surface; avid bow-rider and readily approaches people in the water in some parts of the world.
Distribution: warm temperate, sub-tropical and tropical waters in the N. and S. Atlantic.

Dive sequence   High breaching   Spyhopping   Fast swimming

Striped Dolphin

The ancient Greeks marvelled at the beauty of striped dolphins and painted them in their frescoes several thousand years ago. The various stripes and brush strokes on their bodies, and the bright pink undersides of some individuals, make them look as if they have been hand-painted in real life. They are highly conspicuous animals, seeming to spend an inordinate amount of their time in the air. Their acrobatic repertoire includes breaches up to 7m (23ft) high, belly-flops, back somersaults, tail spins, and upside-down leaps. Their distinctive stripes and high-speed swimming prompted fishermen in the Pacific to call them 'streakers'; the name is also appropriate because they are more easily alarmed than some other dolphins and will turn tail and streak away in some parts of the world.

At a glance
Alternative names: Streaker Porpoise, Euphrosyne Dolphin, Whitebelly, Meyen's Dolphin, Blue-white Dolphin, Gray's Dolphin.
Scientific name: *Stenella coeruleoalba*.
Adult size: 1.8–2.5m (6–8½ft); 90–150kg (200–330lb); males slightly larger than females.
Diet: small squid, fish and crustaceans; Lanternfish is a particular favourite in some areas.
Behaviour: highly conspicuous and capable of amazing acrobatics; will bow-ride.
Distribution: worldwide, mainly in sub-tropical and tropical waters but also warm temperate waters in some parts of range; usually encountered offshore, but will also venture into deep water close to land.

Identification

- slender body
- white or pinkish underside
- dark stripe from eye to flipper
- long, dark side stripe from eye patch along sides
- pale brush stroke from head towards dorsal fin
- dark, prominent dorsal fin
- prominent beak

Distribution map

Dive sequence   Fast swimming

Bottlenose Dolphin

Identification

- robust head and body
- subdued grey colouring
- dark dorsal cape (variable)
- lighter underside
- short but distinct beak
- rounded forehead with marked crease at beak
- prominent, falcate dorsal fin

In many ways, the Bottlenose Dolphin is the archetypal dolphin. The star of films and marine parks, it is the species most people imagine when the word 'dolphin' is mentioned. There are many varieties, living in different parts of the world; some are almost twice as long as others and they vary considerably in shape and colour. Two main types are recognized: a smaller, inshore form found in bays, lagoons and estuaries and a larger, more robust form that lives mainly offshore in waters over or beyond the continental shelf. Sometimes, wild, lone individuals (usually males) become 'friendlies', and seem more interested in sharing the company of human swimmers and small boats than others of their own kind, often remaining in the same area for years.

At a glance

Alternative names: Bottle-nosed Dolphin, Atlantic (or Pacific) Bottlenose Dolphin, Cowfish, Grey Porpoise, Black Porpoise.
Scientific name: *Tursiops truncatus.*
Adult size: 1.9–3.9m (6½–12½ft); 150–650kg (330–1,435lb).
Diet: opportunistic feeder, with a wide range of feeding techniques adapted to local conditions; will take fish, squid, crustaceans and a variety of other prey.
Behaviour: highly active at the surface and frequently lobtails, bow-rides, wake-rides, body-surfs and breaches.
Distribution: cool temperate to tropical waters worldwide; both inshore and offshore. Found in enclosed seas, including the Black and Mediterranean.

Distribution map

Dive sequence

Spyhopping   Breaching

Fast swimming

Risso's Dolphin

Risso's Dolphins are unmistakable, with their slightly bulging foreheads, tall dorsal fins and distinctly battered appearance. The scratches and scars all over the bodies of older animals are mainly caused by the teeth of other Risso's Dolphins, possibly when they are fighting or playing with each other, but confrontations with squid may also be to blame. Their body colour tends to lighten with age and, although there is a great deal of variation between individuals, this adds to the scarring to make some of them almost as white as Belugas. The dorsal fin, flippers and flukes tend to remain darker than the rest of the body. There is a distinctive crease down the centre of the forehead (rarely visible and only then at close range), which is unique to this species. Larger than any other cetacean that carries the name 'dolphin', Risso's Dolphins are fairly common but relatively little studied compared with many of their relatives.

Identification

- robust body extensively scarred
- older animals may be white
- large, rounded head
- indistinct beak
- very tall dorsal fin
- long, pointed flippers

At a glance

Alternative names: Grey Dolphin, Grey Grampus, Grampus, White-head Grampus.
Scientific name: *Grampus griseus*.
Adult size: 2.6–3.8m (8½–12½ft); 300–500kg (660–1,100lb).
Diet: favourite prey is squid, but will also take octopuses, fish and crustaceans.
Behaviour: occasionally does half-breaches and spyhops; will bow-ride but more likely to swim alongside a vessel, or in its wake.
Distribution: cold temperate to tropical waters worldwide; mainly deep offshore waters, or around oceanic islands, but some populations probably resident in shallow coastal waters.

Distribution map

Dive sequence

Spyhopping   Breaching

55

White-beaked Dolphin

Identification

- very robust body
- complex white, grey and black markings
- pale area on tailstock
- white stripe on each side
- very prominent dorsal fin
- short and thick white, brown or grey beak
- dark flippers, dorsal fin and flukes

Strikingly large and robust for a dolphin, the White-beaked has a rather misleading common name. Its beak is not always white. Genuinely white-beaked individuals, which tend to occur more commonly in Europe and live in smaller schools, are distinctive at close range. But western animals generally have grey or even black beaks and live in larger schools (though there are exceptions). Capable of swimming at considerable speed, sometimes churning the water and creating a 'rooster tail' reminiscent of a Dall's Porpoise, the White-beaked Dolphin can be quite acrobatic and will often breach. It shares most of its North Atlantic range with the Atlantic White-sided Dolphin, but ventures farther north into sub-Arctic waters and has the most northerly range of all the world's dolphins.

At a glance

Alternative names: White-beaked Porpoise, White-nosed Dolphin, Squidhound.
Scientific name: *Lagenorhynchus albirostris.*
Adult size: 2.5–3m (8½–10ft); 180–275kg (395–605lb).
Diet: Cod, Herring, Mackerel and variety of other fish, as well as squid, octopus and various crustaceans.
Behaviour: fast, powerful swimmer and sometimes acrobatic; may bow-ride, especially in front of large, fast-moving vessels, although some populations are elusive.
Distribution: widely distributed in cool temperate and sub-Arctic waters of the N. Atlantic, even at the polar ice edge; mostly offshore, but also inshore.

Dive sequence   Breaching   Fast swimming

Atlantic White-sided Dolphin

Atlantic White-sided Dolphins were well known to early fishermen and whalers in the North Atlantic, and they were given names such as 'Springer' or 'Jumper' to reflect their acrobatic nature. Their most distinctive feature is the striking yellow or tan band along each side of the tailstock, although, at first glance, this can make them look superficially similar to Common Dolphins. However, the position of the band, the complex grey, white and black body markings with no distinctive hourglass pattern, the stockier body and the shorter beak make the Atlantic White-sided Dolphin quite a distinctive species at close range. It often feeds in association with large whales such as Fins and Humpbacks, perhaps chasing similar prey, and seems to enjoy playing around them – sometimes even riding in their bow-waves.

At a glance

Alternative names: Lag, Jumper, Springer, Atlantic White-sided Porpoise.
Scientific name: *Lagenorhynchus acutus*.
Adult size: 1.9–2.5m (6½–8½ft); 165–200kg (365–440lb).
Diet: variety of fish, including Cod, Herring and young Mackerel, as well as squid and crustaceans.
Behaviour: acrobatic and fast swimmer; frequently bow-rides in front of fast vessels, although can be wary of boats and ships in some areas.
Distribution: cold temperate and sub-Arctic waters of the northern N. Atlantic, especially along the continental shelf edge.

Identification

- robust body with thick tailstock
- black or dark grey upperside
- grey stripe along flanks
- white patch below dorsal fin
- yellowish patch on tailstock
- white underside
- short, thick beak
- tall, falcate dorsal fin

Distribution map

Dive sequence

Fast swimming

Fraser's Dolphin

Identification

- stocky build
- blue-grey or grey-brown upperside
- creamy white or pinkish underside
- dark lateral stripe (variable width and intensity and sometimes absent)
- short but well-defined beak
- small dorsal fin
- tiny flippers

Distribution map

Fraser's Dolphin was not scientifically described until 1956, after cetologist Francis Charles Fraser found a mislabelled skeleton in the British Museum that had been collected 60 years earlier from a beach in Sarawak, Malaysia. No-one knew what the animal looked like in real life, or where it lived, until the early 1970s when several complete specimens were stranded in widely separated parts of the world. The species was first seen alive at around the same time. Fraser's Dolphins have been observed on many occasions since then and do not appear to be as rare as was once thought, but they remain poorly known. Most schools are large, containing 100 to 500 and as many as 1,000 individuals, and are often associated with other species of warm water toothed whales and dolphins.

At a glance

Alternative names: Fraser's Porpoise, White-bellied Dolphin, Sarawak Dolphin, Shortsnout Dolphin, Bornean Dolphin.
Scientific name: *Lagenodelphis hosei.*
Adult size: 2–2.6m (6½–8½ft); c.160–210kg (350–460lb).
Diet: mid-water fish, squid and crustaceans.
Behaviour: aggressive swimming style, often leaving the water in a burst of spray; will bow-ride in many parts of its range.
Distribution: deep warm temperate to tropical waters worldwide, although relatively scarce in the Atlantic and very few records in Europe; rarely seen inshore, except around oceanic islands and in areas with narrow continental shelf.

Dive sequence

Fast swimming

Harbour Porpoise

The Harbour Porpoise is the only porpoise living in Europe. It is commonly seen along many coasts, although it often drowns in fishing nets and, consequently, has disappeared from some parts of its former range. It can be difficult to observe closely and is generally wary of boats – a brief glimpse of its dark back and low, triangular dorsal fin is all this undemonstrative little cetacean usually shows of itself. When it rises to breathe, the lasting impression is of a slow, forward-rolling motion, as if the dorsal fin is mounted on a revolving wheel lifted briefly above the surface and then withdrawn. It can move surprisingly fast as well, and as it splashes to the surface it is almost impossible to see what is happening. While some whalewatchers find it a frustrating animal to observe, most come to enjoy the familiar sneezing sound of its invisible blow; this gives the Harbour Porpoise one of its alternative names, the Puffing Pig. It is believed to be one of the shortest-lived cetaceans, rarely surviving past the age of 12 years.

At a glance
Alternative names: Common Porpoise, Puffing Pig.
Scientific name: *Phocoena phocoena*.
Adult size: 1.4–1.9m (4½–6½ft); 55–65kg (125–145lb).
Diet: mainly schooling, non-spiny fish such as Herring, Whiting and Mackerel, but also other fish, some squid and possibly crustaceans.
Behaviour: slow swimming behaviour very distinctive, but may also swim fast and produce considerable spray; generally uninterested in boats.
Distribution: cool temperate and sub-Arctic coastal waters of northern hemisphere.

Identification

- nondescript colouring (dark above, light below)
- small size and robust shape
- low, triangular dorsal fin
- small, rounded head
- no forehead or distinct beak
- slow, forward-rolling motion

Distribution map

Dive sequence

WHERE TO WATCH WHALES IN EUROPE

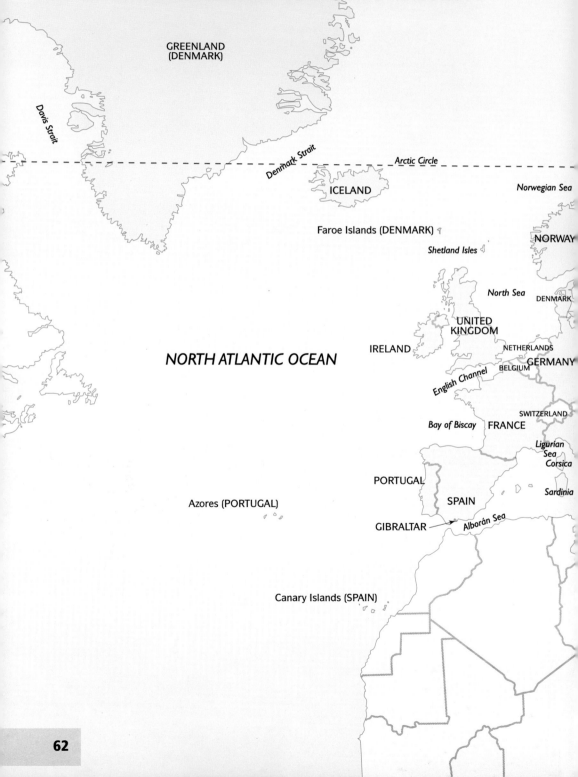

MAP OF WHALEWATCHING COUNTRIES IN EUROPE

Greenland Sea

GREENLAND
(DENMARK)

Davis Strait

Denmark Strait

Arctic Circle

ICELAND

Norwegian Sea

Faroe Islands (DENMARK)

Shetland Isles

NORWAY

North Sea

DENMARK

UNITED
KINGDOM

IRELAND

NETHERLANDS

BELGIUM

GERMANY

NORTH ATLANTIC OCEAN

English Channel

Bay of Biscay

FRANCE

SWITZERLAND

Ligurian
Sea

Corsica

PORTUGAL

SPAIN

Sardinia

Azores (PORTUGAL)

GIBRALTAR

Alborán Sea

Canary Islands (SPAIN)

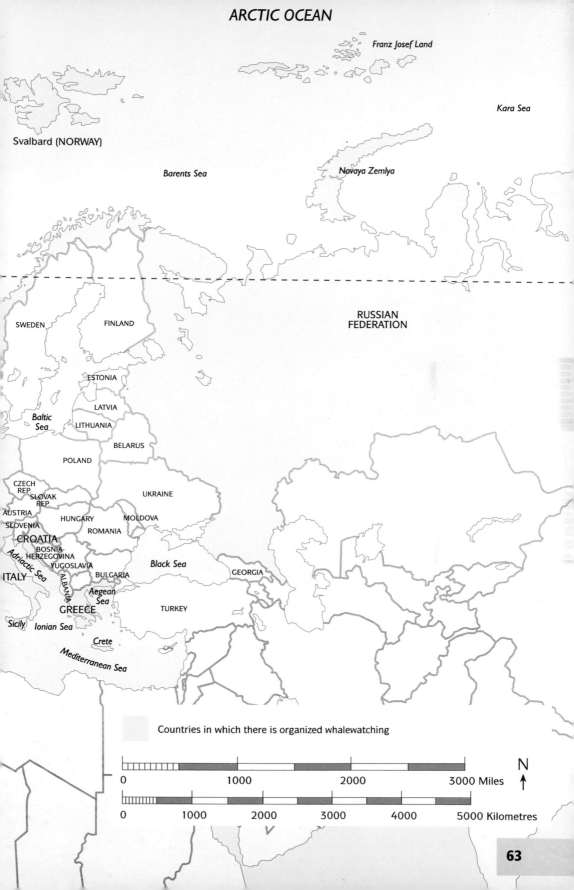

ARCTIC OCEAN

Franz Josef Land

Kara Sea

Svalbard (NORWAY)

Barents Sea

Novaya Zemlya

RUSSIAN
FEDERATION

SWEDEN

FINLAND

ESTONIA

LATVIA

Baltic
Sea

LITHUANIA

BELARUS

POLAND

CZECH
REP.

SLOVAK
REP.

UKRAINE

AUSTRIA

SLOVENIA

HUNGARY

MOLDOVA

CROATIA

ROMANIA

BOSNIA-
HERZEGOVINA

YUGOSLAVIA

Adriatic Sea

Black Sea

GEORGIA

ITALY

ALBANIA

BULGARIA

Aegean
Sea

GREECE

TURKEY

Sicily

Ionian Sea

Crete

Mediterranean Sea

Countries in which there is organized whalewatching

| 0 | | 1000 | | 2000 | | 3000 Miles |

| 0 | 1000 | 2000 | 3000 | 4000 | 5000 Kilometres |

N

THE AZORES (Portugal)

Main species: Sperm Whale, Short-finned Pilot Whale, various beaked whales, Short-beaked Common Dolphin, Atlantic Spotted Dolphin, Striped Dolphin, Bottlenose Dolphin, Risso's Dolphin; increasingly regular sightings of Blue Whale, Fin Whale and Sei Whale.

Main locations: most operators based in Horta (Faial) or Lajes (Pico), but trips also available from Flores, São Miguel and other islands.

Types of tours: half-day tours, extended multi-day expeditions, research programmes, and land-based observation points.

When to go: best weather and sea conditions May to October (but many species are present year-round, including Short-beaked Common, Bottlenose and Risso's Dolphins and Short-finned Pilot Whale); male Sperm Whales year-round, females and calves most abundant May to October; Blue, Fin and Sei Whales mainly spring; Atlantic Spotted Dolphins mainly end of June to end of October.

Contact details: Aquaaçores: www.aquaacores.com
Espaço Talassa: www.espacotalassa.com
Pico Sport: www.whales-dolphins.net
Whale Watch Azores: www.whalewatchazores.com
Horta Cetáceos; informacoes@hortacetaceos.com
Rotas das Baleias: www.espacotalassa.com
Futurismo Azores Whale Watching: www.ciberacores.com/futurismo
Norberto Diver: www.norbertodiver.com
Searide Açores: www.searideacores.pt/
Nauticorvo: www.nauticorvo.pt/
Gaspar Ocean Adventures: www.gaspar-ocean-adventures.com

Outside the Azores: Whale and Dolphin Conservation Society: www.bluetravel.co.uk
Discover the World: www.arctic-discover.co.uk
WildOceans: www.wildwings.co.uk/wildoceansintro.html
Colibri Umwelt Reisen: www.colibri-berlin.de
Atalante: www.atalante.fr
Terra Incognita: www.terra-incognita.fr

Imagine floating alongside a pod of Sperm Whales in the shadow of the impressive volcanic cone of Pico – the highest point in the Azores – or watching beaked whales from a purpose-built observation tower with sweeping views over this remote corner of the North Atlantic. The isolated archipelago of the Azores has a well-deserved

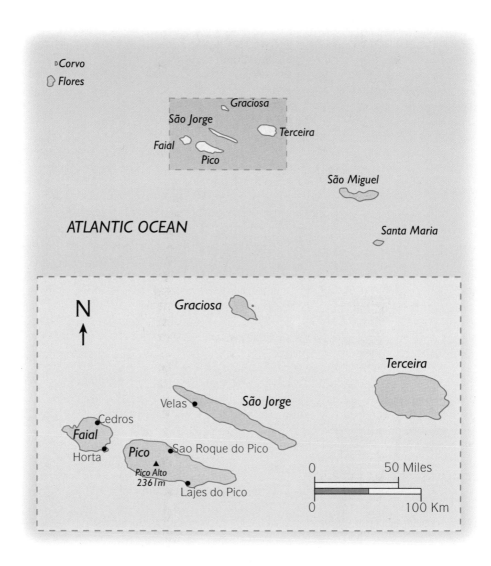

reputation as one of the best whalewatching destinations
in the world. Lying some 1,400-1,800km (870-1,120 miles)
west of Portugal, and about a third of the way across the
Atlantic towards North America, it is the permanent or
temporary home of more than a quarter of all known
cetacean species.

There are nine islands altogether and, according to legend,
they are all that remain of the lost continent of Atlantis.
Volcanic in origin, they rise dramatically out of the Mid-
Atlantic Ridge (where the American, Eurasian and African

Below: *Open sailboats, or canoas, were once used to hunt Sperm Whales* (Physeter macrocephalus) *around the Azores and, nowadays, can be seen being raced in local festivals.*

continental plates meet) and form three widely separated groups stretching over several hundred kilometres of ocean. Discovered in 1427 by Portuguese explorers, and colonized shortly afterwards, they have been Portuguese territory ever since.

Whaling

Whales have played an integral role in Azorean life for many years and, until quite recently, the focus of attention

was on whaling. The first Yankee whalers arrived in the islands in 1765 and the locals were soon being employed on the American whaling ships. As Herman Melville explains in *Moby Dick*, 'no small number of these whaling seamen belong to the Azores, where the outward bound Nantucket whalers frequently touch to augment their crews from the hardy peasants of those rocky shores.' Whaling became an important part of the economy and the Azoreans quickly developed a reputation for being both brave and tough.

Ultimately, they designed and built their own whaling vessels. Longer, narrower and faster than the American boats that were their inspiration, they were dubbed the 'Pico arrows'. The whales were spotted from lookout towers, called vigias, which were built on prominent headlands overlooking the sea. As soon as a Sperm Whale came into view, the spotter shouted *'baleia à vista'* ('whale in sight') and fired a rocket into the air as a signal. Like volunteer lifeboatmen, the whalers rushed down to the boathouses and lowered their boats into the water. They used open sailboats, or canoes, and followed directions from the spotter on shore (who signalled with sheets arranged on the ground) to manoeuvre close enough for the harpooner to hurl a barbed harpoon into the whale's back. This did not kill the animal, but merely attached it to the boat. Then it was a dangerous waiting game, until the whale tired and the harpooner could dispatch it with repeated thrusts of a lance. Typically, it took an hour to kill an adult Sperm Whale, although sometimes the battle for life or death took an entire day.

Even after the rest of the world had progressed to steam-driven catcher boats and explosive harpoons, and long after the last American whaler had disappeared from the North Atlantic, the Azoreans continued to hunt using this traditional method. The spotter swapped his sheets for a radio, and motorboats were used to tow the canoes out to the whales and then to retrieve the carcasses, but otherwise little changed until the last whale was killed in the 1980s. The hunt officially stopped in 1984, although three additional whales were taken in 1987. There were several factors in the demise of the industry: the whalers found less demanding alternative employment; there was a lack of legitimate markets for whale oil; and the owner of the last working whaling factory died.

Although it had been conducted throughout the archipelago, whaling was most significant on the island of Pico and it is still possible to see some of the remnants and relics of the industry. The Museu dos Baleeiros (Whalers' Museum) in Lajes do Pico is open year-round and includes the three original nineteenth-century boathouses. The last whaling factory to be used in the Azores, on the other side of Pico, can also be explored. In addition, the Museum of the Sea will soon be opening at the site of the whaling factory in Horta, on the island of Faial, and there is a scrimshaw museum above Peter Café Sport, also in Horta. It is possible to watch the original canoas being raced in local festivals throughout summer.

Whalewatching

Commercial whalewatching in the Azores began in 1991 – just four years after the last whales had been killed. It has grown exponentially since then and, already, nearly 15,000 whalewatchers migrate to the archipelago every year. There are many different tours and vessels, ranging from three-hour trips on rigid-hulled inflatables to week-long expeditions on traditional sailing schooners. Some operators encourage participants to get involved with research – by helping to track Sperm Whale movements, filling in data logs, collecting skin samples for DNA analysis, and so on. There is also an excellent three-hour walk, in Lajes do Pico, clearly marked and designed specifically for whalewatchers, which includes some of the best vantage points and several key sites from the old whaling days.

Whalewatching in the Azores is very special, because there is an extraordinary and very effective partnership between the old tradition of whaling and the new industry of whalewatching. The original vigias have been restored by whalewatch companies and they are now manned by ex-whalers who radio sightings to the boat skippers. In the old days, the whalers were only interested in Sperm Whales (all other cetaceans were just 'fish') but nowadays they apply their keen eyesight and expert spotting skills to a much wider range of subjects. Meanwhile, they are training young Azoreans who will take over as the land-based spotters in years to come. Ex-whalers are also being employed by some operators as skippers and crew.

Opposite: A rare but memorable encounter with a group of Northern Bottlenose Whales (Hyperoodon ampullatus), photographed in the Azores. Note their bulbous foreheads. They are quite inquisitive and may approach stationary boats.

Many of the vigias, especially on the islands of Faial and Pico, are open to visitors. One of the best is the Vigia da Queimada, just 1km (0.6 miles) from Lajes do Pico, which has a sweeping view of some 200 degrees from east to west and makes such a good vantage point that on a clear day it is possible to spot whales up to 30km (19 miles) away. Built in 1936, it was the last vigia to be used by whalers in the mid-1980s and was renovated in 1991. It is unusual in having two levels – the upper level is for professional spotters directing whalewatch boats to the whales and the lower level is for visitors. Another vigia, Vigia do Arrife, about 2km (1¼miles) from Lajes, is also very popular. However, whalewatching from the vigias is not easy. Unless you are an experienced observer, it can be difficult to spot whales and dolphins over such vast distances, let alone identify them.

Whalewatch permits are issued by the Department of Tourism and regulations are being introduced. These include: only one boat is allowed with a whale or a group of whales at any one time; the first boat can remain with the whales for 10-15 minutes, then has to let another boat have access; boats must never approach from the front or sides; and boats can only approach to within 50m (165ft) or 100m (330ft) if a calf is present. Many operators are highly professional and responsible, but there is an urgent need for some to improve their behaviour around the whales and to make their trips more educational.

Where to go

Whalewatching in the Azores has at least a 95 per cent success rate in finding whales or dolphins. It is possible to encounter more than half a dozen different species in a single trip (two or three is more typical) and no fewer than 24 species have been recorded altogether.

There are two major hotspots – off the south coast of Pico and off the north coast of Faial – but there are many other excellent areas to watch cetaceans around the archipelago. Sperm Whales are the most frequently seen species here and, indeed, groups of females and their calves can be seen almost anywhere in the Azores. Many operators have onboard hydrophones to help them locate these extraordinary whales.

Short-finned Pilot Whales, Short-beaked Common Dolphins, Bottlenose Dolphins, Atlantic Spotted Dolphins and Risso's Dolphins are also very common. Striped Dolphins are encountered fairly regularly and Blue, Fin and Sei whales turn up increasingly regularly, especially from late March to June.

The Azores is also good for beaked whales. It is not unusual to see several different species from shore, although boat-based encounters are relatively infrequent. Espaço Talassa, the largest whalewatch operator in the Azores, estimates that Sowerby's Beaked Whales are seen on approximately one in every 20 trips. Cuvier's Beaked Whales are often seen from the vigias, and are probably quite common in the Azores, although they are rarely observed from the boats. There have been some wonderful encounters with Northern Bottlenose Whales in recent years but these are few and far between. True's Beaked Whale has been positively identified by Lisa Steiner (Whale Watch Azores) and Gervais' Beaked Whales have also been recorded. Several other species have been encountered, albeit infrequently, and these include Humpback Whale, Minke Whale, Pygmy Sperm Whale, Dwarf Sperm Whale, Killer Whale and Rough-toothed Dolphin.

Historically, Northern Right Whales have also been recorded (but after centuries of whaling they are close to extinction and no more than a handful survive in the north-east Atlantic). With more and more people looking out to sea, there is every chance that new species will be added to the list in the future – after all, the Azores is one of those places where almost anything can turn up.

Swimming with whales

Swimming with whales is now prohibited under new legislation (except under special licence for scientific or professional filming work) but some operators still take people out to swim with dolphins. However, the dolphins are not habituated to people and generally do not interact with swimmers.

BAY OF BISCAY

Main species: Minke Whale, Sei Whale, Fin Whale, Sperm Whale, Northern Bottlenose Whale, Cuvier's Beaked Whale, Long-finned Pilot Whale, Killer Whale, False Killer Whale, Short-beaked Common Dolphin, Striped Dolphin, Bottlenose Dolphin, Risso's Dolphin, Harbour Porpoise.

Main locations: cruise-ferry routes: *Pride of Bilbao* (operated by P&O Portsmouth) makes two return trips each week year-round between Portsmouth and Bilbao; *Val de Loire* (operated by Brittany Ferries) makes two return trips each week from April to November between Plymouth and Santander.

Types of tours: cruise-ferries travelling between southern England and northern Spain (a 'mini-cruise' of just under three days on the *Val de Loire* and four days on the *Pride of Bilbao* includes a few hours ashore between the outbound and return journeys). From summer 2002, the Biscay Dolphin Research Programme will begin dedicated research trips aboard the *Loyal Helper* with limited availability (ten per trip) for paying passengers.

When to go: cetaceans can be observed in the Bay year-round, although the variety of species and their relative abundance vary greatly from month to month; the best period for both weather and whalewatching tends to be summer and autumn (June–September).

Contact details: *To book direct:* P&O Portsmouth Ferries: www.poportsmouth.com
Brittany Ferries: www.brittany-ferries.com
To join organized research and whalewatching tours:
Biscay Dolphin Research Programme: www.biscay-dolphin.org.uk
ORCA: www.orcaweb.org
The Company of Whales: www.companyofwhales.co.uk
Naturetrek: www.naturetrek.co.uk
WildOceans: www.wildwings.co.uk

Bordered by the Atlantic coasts of France and Spain, the Bay of Biscay was largely overlooked as a whalewatching destination until surprisingly recently. Now it has earned a well-deserved reputation as one of the most exciting, accessible and affordable places in Europe to watch an impressive variety of whales, dolphins and porpoises.

More than a dozen species are seen regularly and a number of others have been recorded at one time or another. Sowerby's Beaked Whales and Atlantic White-sided Dolphins have been encountered in recent years, and the first confirmed live sighting of True's Beaked

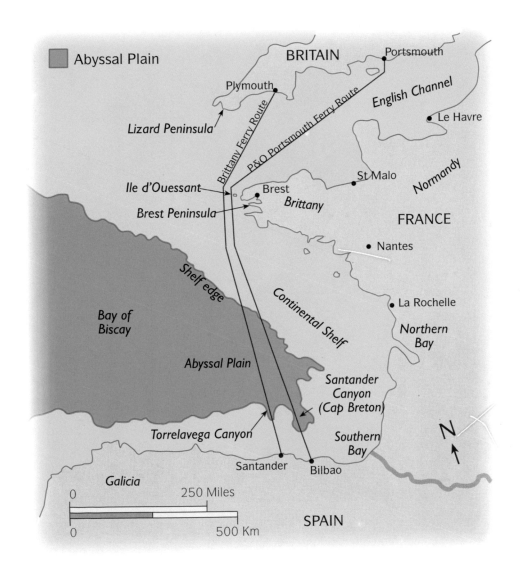

Whale in the eastern North Atlantic, and possibly anywhere in the world, was made by ORCA surveyors on 9 July 2001 in the southern part of the Bay. Blue and Humpback Whales are occasionally recorded; Pygmy Killer Whales have been seen in the Bay three times (the first records for northern Europe); and individual Blainville's and Gervais' Beaked Whales, as well as Dwarf Sperm Whales, Pygmy Sperm Whales, Fraser's Dolphins and Rough-toothed Dolphins, have all stranded in the region but are yet to be positively recorded at sea.

Northern Right Whales once occurred in large numbers, but centuries of whaling wiped out all but a handful of the population in the eastern North Atlantic. Basque whalers, from France and Spain, were probably the world's first commercial whalers and, by about 1200, were using small boats, called chalupas, to hunt Right Whales in the Bay. It was a highly dangerous occupation as a prayer the whalers recited just before harpooning a whale indicates: 'Allow us, Mighty Lord, to quickly kill the great fish of the sea; without injuring any one of us when he is bound by the line in his tail or his breast; without tossing the boat's keel skyward, or pulling us with him to the depths of the sea; the profit is great, the peril is also great; guard above all our lives.'

Cruise ferries

Most of what we know about cetaceans in the Bay of Biscay today has come from the work of two independent voluntary research organisations working on passenger cruise-ferries plying the waters between southern England and northern Spain: the Biscay Dolphin Research Programme (BDRP) and Organisation Cetacea (ORCA). BRDP is supported by P&O Portsmouth, the Sea Watch Foundation and the Spanish Cetacean Society, and has been conducting year-round monthly surveys aboard the Portsmouth-Bilbao ferry *Pride of Bilbao* since 1995. ORCA is supported by the Whale and Dolphin Conservation Society, English Nature and P&O Portsmouth, and has been conducting similar surveys aboard *Pride of Bilbao*, the Plymouth–Santander ferry *Val de Loire* and several other vessels since 1997.

The ferries are extremely comfortable (more like cruise ships than passenger ferries) with swimming pools, saunas, gyms, cinemas and other facilities. More importantly, they are so large and stable that they allow the use of tripod-mounted telescopes in all but the roughest seas. The best watching on both vessels is from the upper deck at a height of about 30m (100ft) above sea level but, on colder days, limited viewing is possible from the more comfortable indoor lounges.

It is possible to book direct with either P&O Portsmouth or Brittany Ferries, or to join themed cruises run in conjunction with either of the two research groups; researchers give lectures and illustrated slide shows and will help passengers to find and identify any cetaceans encountered.

Opposite: Short-beaked Common Dolphins (Delphinus delphis) photographed from a passenger cruise-ferry in the Bay of Biscay, as they swim fast alongside. Notice the distinctive hourglass pattern on their sides.

Pride of Bilbao has an enormous interpretation board with text and illustrations, to help with identification, and a dedicated observatory on Deck 8; BDRP has also introduced a full-time onboard wildlife officer to give lectures and assist whalewatchers throughout the summer. Three new education, interpretation and identification boards, covering cetaceans, seabirds and other marine wildlife, are due to be erected on the *Val de Loire*.

There are some subtle differences in whalewatching from the two ferries, but these are more to do with sailing times than routings. There is little agreement about which is best – it all depends on whether you want to see beaked whales or baleen whales and, consequently, whether you spend more daylight hours over the continental shelf or the submarine canyons. Apart from the early stages of the outbound journey (*Val de Loire* sails due south from Plymouth across the Western Approaches while *Pride of Bilbao* sails south-west through the English Channel) the routes broadly coincide from Isle d'Ouessant off Brittany's Brest Peninsula for the remainder of the journey to northern Spain.

Underwater landscape

One of the principal reasons for the variety and abundance of cetaceans observed from the two ferries is the ever-changing underwater landscape en route. From a whalewatching perspective, the journey can be split into four main regions: the English Channel, the Brittany Peninsula section of the continental shelf, the abyssal plain and two submarine canyons.

This part of the English Channel is bordered by England's Lizard Peninsula in the north and France's Brest Peninsula in the south. It consists of shallow seas (typically less than 80m (260ft)), with a few deeper trenches dropping to depths of up to 110m (360ft). In comparison with other parts of the journey, the Channel is not particularly productive for cetaceans. However, Harbour Porpoises are recorded regularly (even though they can be difficult to see from the high observation decks of the two passenger ferries) and have even been encountered just outside Portsmouth Harbour. Small groups of Short-beaked Common Dolphins are occasionally seen here, especially from October to January (there are many more of them in the Bay itself), along with small numbers of Bottlenose Dolphins and occasional Long-finned Pilot Whales.

Opposite: Striped Dolphins (Stenella coeruleoalba) are a familiar sight for whalewatchers over the deep waters of the central Bay of Biscay and along the continental shelf edge. This fast swimmer reveals its pinkish belly.

In recent years, during the summer, Minke Whales have also been encountered regularly in the Channel, often accompanied by calves.

Over the next part of the continental shelf, beyond Isle d'Ouessant, cetacean sightings usually increase. The water here ranges in depth from 110m (360ft) to 200m (655ft) and is home to all four Channel species as well as Risso's Dolphins and, particularly late in the summer, Minke Whales. The edge of the continental shelf is up to 100km (60 miles) south of the Brest Peninsula (the precise distance depends on the crossing point and the angle of slope, which varies from shallow to very steep). Forming a distinctive border around the deep waters of the central Bay, with cold, nutrient-rich upwellings teeming with plankton and fish, this area is excellent for cetaceans and the Channel and continental shelf species are typically joined by Fin Whales and Striped Dolphins. Migratory species may also use the shelf edge as a navigational aid, bringing them into the Bay on their travels north or south.

Within a distance of less than 50km (30 miles) the seabed drops from the shelf edge to depths of more than 4.5km (2.8 miles) in the abyssal plain. The journey traverses a 95km (60 mile) stretch of this deep water before crossing the southern edge of the continental shelf some 95km (60 miles) off the coast of Spain. There tend to be relatively fewer cetacean sightings over the abyssal plain itself, although it is one of the best areas for Striped Dolphins, is excellent for Fin Whales in late summer, and is the home of deep-diving Sperm Whales and beaked whales. Short-beaked Common Dolphins are also very abundant here and large groups often come to ride in the bow waves of the ships.

The southern edge of the continental shelf, less than 75km (45 miles) from the Spanish coast, is bisected by two deep-water trenches or submarine canyons: the Torrelavega and Cap Breton (sometimes called Santander) Canyons. It is hard to imagine their immense scale but, as a guide, they are both more than twice the depth of Arizona's Grand Canyon. The ferries cruise the length of these canyons for about 32km (20 miles) (Cap Breton on the P&O Portsmouth route and Torrelavega on the Brittany Ferries route) before the seabed rises steeply once again towards the northern coast of Spain. This is the area where Northern Bottlenose Whales and Cuvier's Beaked Whales are consistently seen and, indeed, the canyons are currently considered among the best places in the world for encountering these elusive animals (although Northern Bottlenose have not been seen in any numbers over Cap Breton Canyon since 1999).

Below: *False Killer Whales* (Pseudorca crassidens) *are among the many more unusual species occasionally encountered in the Bay of Biscay.*

Whalewatching in the Bay of Biscay is year-round, although it can be a harsh and forbidding place in winter and there are dramatic changes in which species can be seen and their abundance with the seasons. The only cetaceans seen regularly throughout the year are Bottlenose, Short-beaked Common and Striped Dolphins, as well as Long-finned Pilot Whales and even Cuvier's Beaked Whales, but other species occur in impressive numbers during key periods.

Seasonal variations

Long-finned Pilot Whales, Killer Whales and False Killer Whales begin to appear in April and can be encountered throughout the summer until early autumn. Long-finned Pilot Whales are common, while Killer and False Killer Whales tend to occur in small numbers. The first records of Risso's Dolphins have been in April, although they can be encountered at almost any time of year. As spring arrives and the water warms, Short-beaked Common Dolphins begin to move away from the Brest Peninsula and head further south towards the edge of the continental shelf. Meanwhile, Striped Dolphins are doing the reverse by moving progressively north from the lower part of the Bay. Spring is a good time to see migratory whales: Humpback and Sperm Whales are the most commonly encountered, but there are also small numbers of Blue Whales. This is a good time of year for Cuvier's Beaked Whales, which are frequently seen over the canyons in the south.

By May and June, the diversity of cetaceans in the Bay has noticeably increased. Harbour Porpoises, Bottlenose Dolphins and Minke Whales are commonly seen in the Channel at this time and Northern Bottlenose Whales, Cuvier's Beaked Whales and White-beaked Dolphins, to name a few, can be encountered in the Bay itself. Short-beaked Common Dolphins are relatively hard to find this early in the summer, but they reappear in huge numbers in late June or July, when groups of up to 3,500 have been recorded and many of the females are accompanied by calves. The high summer is best for sheer variety: the Bay is teeming with cetaceans during June, July, August and September. On average, there is nearly one sighting per hour at this time of year and the records to date are 105 separate sightings and 13 different species during single trips. Northern Bottlenose Whales are among the many species seen regularly at this time of year.

Opposite: *Whalewatchers keep vigil on the* Pride of Bilbao *as the sun sets over the Bay of Biscay, hoping for a glimpse of whales or dolphins.*

Large baleen whales move into the Bay during late June and July and inhabit the central region over the deep abyssal plain. Blue Whales are seen fairly regularly (although in very small numbers) during the summer and, less regularly, during spring. Fin Whales, in particular, often occur in enormous numbers from August to October and there are often a few Sei Whales among them. Over the submarine canyons, Cuvier's Beaked Whale numbers seem to peak in August and the majority of Northern Bottlenose Whale sightings are in the same month (although this may partially reflect observer bias). The dolphins tend to disperse in August, September and October, although they still occur in large numbers.

Record-breaking crossing

On one memorable crossing, BDRP observers on the *Pride of Bilbao* travelling at 20 knots took 40 minutes to pass through a single group of Fin Whales extending over 365 sq km (140 sq miles).

As autumn approaches, the spring trends reverse. Dolphin groups split up, Striped Dolphins begin to return south and Short-beaked Common Dolphins head back north. The large whales begin to gather over the southern half of the abyssal plain and, by the last week in September or the first week in October, the last huge gatherings of them are dispersing. Migrants pass through on their way back south during October and then the great whales are gone. There are no more than occasional sightings of them throughout autumn and winter. By late October, the Striped Dolphins that are such a feature of summer have all but disappeared and Cuvier's Beaked Whales take over the submarine canyons once again.

Many calm winter crossings are rewarded with small groups of 5–50 Short-beaked Common Dolphins, especially around the edge of the continental shelf off the Brest Peninsula, and Long-finned Pilot, Minke and Fin Whales are occasionally seen. Sowerby's Beaked Whales have also been recorded during winter and, altogether, no fewer than 11 species have been seen at this time of year.

Other wildlife

The ferry crossings offer the chance to see a lot of other wildlife, as well. Seabirds are particularly prominent, with Gannets, Kittiwakes, Fulmars, Great Skuas, Guillemots and Razorbills being the most regular. In deeper water, it is worth scanning around fishing boats for a wonderful range of possible species including Little and Sabine's Gulls, Manx, Sooty and Great Shearwaters, Arctic and Pomarine Skuas, and Wilson's Storm Petrels. The deeper waters of the central and southern parts of the Bay are not populated by large numbers of birds, but observers are sometimes rewarded with specialities such as Mediterranean Shearwaters.

During the spring and autumn migrations, it is not unusual to encounter a variety of land birds travelling between their summer and winter haunts. These include finches, warblers and wagtails as well as birds of prey such as Hobbies, Kestrels, Montagu's Harriers and Merlins; some resourceful birds have even been known to hitch a ferry-ride virtually all the way across the Bay. Several species of large fish can also be seen, especially Basking Sharks, but also Thresher and Porbeagle Sharks. The occasional Ocean Sunfish or Leatherback Turtle may drift by and it is not unusual to see Atlantic Flying Fish.

Beaked whales in the Bay of Biscay

The beaked whale family includes some of the least known of all the world's cetaceans. Much of the information we have about them has been gleaned from dead animals washed ashore and, until surprisingly recently, few had been observed properly at sea. The Bay of Biscay, however, is rapidly gaining a reputation as a hotspot for several members of the family. Cuvier's Beaked Whale and Northern Bottlenose Whale are now being encountered on a regular basis and both Sowerby's and True's Beaked Whales have been positively identified in the Bay in recent years. There may even be potential for studying Blainville's and Gervais' Beaked Whales, which have both stranded in the region but are yet to be positively identified at sea.

BRITAIN

Main species: Minke Whale, Long-finned Pilot Whale, Killer Whale, Short-beaked Common Dolphin, Bottlenose Dolphin, Risso's Dolphin, White-beaked Dolphin, Atlantic White-sided Dolphin, Harbour Porpoise.

Main locations: Cornwall, Dorset (England); Pembrokeshire, Ceredigion (Wales); Isle of Man; Channel Islands; Inner and Outer Hebrides, Northern Isles, Highland (Scotland).

Types of tours: half- and full-day tours, extended multi-day expeditions, research programmes, and land-based observation points.

When to go: resident Bottlenose Dolphin populations and Harbour Porpoises present year-round; most other species inshore mainly April to October; prime season for most species and best weather conditions May to early October.

Contact details: *Scotland:* Hebridean Whale and Dolphin Trust: www.hwdt.org
Sea Life Surveys: www.sealifesurveys.co.uk
Speyside Wildlife: www.speysidewildlife.co.uk
Western Isles Sailing & Exploration Company Ltd: www.wisex.co.uk
Sea.fari Adventures: www.seafari.co.uk/skye
Arisaig Marine: www.arisaig.co.uk
Northern Light: www.northernlight-uk.com
NorthCoast Marine Adventures: www.northcoast-marine-adventures.co.uk
Strond Wildlife Charters: www.erica.demon.co.uk/Strond.html
Wild Country Expeditions: www.outdoor-scotland.co.uk
Inter-Island Cruises: www.jenny.mull.com; www.whalewatchingtrips.co.uk
Minch Charters: www.scotland-inverness.co.uk/sulair.htm
Dolphin Ecosse: www.dolphinecosse.co.uk
Ardnamurchan Charters: www.west-scotland-tourism.com/ ardnamurchan-charters/
Dolphins & Seals of the Moray Firth Visitor & Research Centre:
www.highland.gov.uk/educ/publicservices/visitorcentres/dolphins.htm
Gairloch Marine Life Centre and Sail Gairloch Cruises: www.porpoise-gairloch.co.uk
Friends of the Moray Firth Dolphins: www.loupers.com
Moray Firth Wildlife Centre: www.mfwc.co.uk
Dolphin Trips Avoch: www.dolphintripsavoch.co.uk
Moray Firth Cruises: www.users.globalnet.co.uk/~nkhtl/cruise.html
Benbola Wildlife Cruises: www.mfwc.co.uk/boatrips.htm
Wales: Cardigan Bay Marine Wildlife Centre: www.marine.wildlife.centre.freeservers.com
England: Seawatch Charter: www.whaleguide.com/directory/seawatchcharter.htm
Durlston Marine Project: www.durlston.co.uk/marine
Isle of Man: The Basking Shark Society: www.isle-of-man.com/interests/shark
General: Sea Watch Foundation: www.seawatchfoundation.org.uk
Whale and Dolphin Conservation Society: www.wdcs.org
International Fund for Animal Welfare: www.ifaw.org
The Wildlife Trusts: www.wildlifetrusts.org

No fewer than 27 different cetaceans (nearly a third of all the world's species) have been recorded in the waters around England, Scotland and Wales over the years.

Some have been encountered just once, or only a handful of times. These include Sowerby's, Cuvier's, True's and Blainville's Beaked Whales, as well as False Killer Whale, Melon-headed Whale, Pygmy Sperm Whale, Northern Right Whale, and even Beluga and Narwhal. Many others are seen quite frequently and, in some cases, even on a daily basis.

Above: *Already fairly common in offshore waters in the north-west of Britain, Atlantic White-sided Dolphins* (Lagenorhynchus acutus) *seem to be extending their range into the North Sea.*

The most commonly sighted species tend to be the smaller whales, dolphins and porpoises. They are all encountered more frequently along the Atlantic seaboard than anywhere else, but can be seen in other parts of Britain as well. Most of them appear to have seasonal inshore–offshore migrations and, with a few exceptions, are more frequently encountered close to shore during the height of summer. At other times, they tend to be further offshore or further south.

The Harbour Porpoise is perhaps the most familiar species, mainly because of its predominantly inshore distribution, and is unusual in occurring around virtually the entire British coast. It tends to be more common in the north and west (the population may have declined in the south in recent years) and is the only species seen regularly close to shore in the North Sea.

There are three known resident populations of Bottlenose Dolphins: in the Moray Firth, Scotland; Cardigan Bay, Wales; and in the waters off Dorset, Devon and Cornwall, England (a nomadic group, which divides its time between the waters of these three counties). Elsewhere, Bottlenose Dolphins are seen most often in the north and west and it is possible that future research may identify additional resident groups in these areas.

Four other dolphin species are seen with some regularity: Risso's, which is widely distributed along the west coast and is particularly common around the Isle of Man and off the Eye Peninsula on the Isle of Lewis as well as off Bardsey Island, north-west Wales; White-beaked, which is quite common along the northern Atlantic seaboard and in the central and northern parts of the North Sea, and moves inshore during late summer and autumn; Atlantic White-sided, which is fairly common in offshore waters in the north-west but seems to be extending its range into the northern part of the North Sea, and also moves inshore in late summer and autumn; and the Short-beaked Common Dolphin, which occurs mainly in the south-west in the western approaches to the English Channel (although it seems to be extending its range northwards into Scotland and occasionally enters the North Sea). Striped Dolphins are seen very occasionally, but spend most of their time in offshore waters to the south-west.

Friendly dolphins

Britain has had a number of friendly Bottlenose Dolphins over the years: solitary individuals named Freddy, Donald, Percy, Simo and Georges have all befriended people and taken up residence close to coastal towns for as long as several years.

Above: *The Blue Whale (Balaenoptera musculus), the largest animal on the planet, has been reported far off the north-west coast of Scotland.*

Killer Whales are seen irregularly and in small numbers, mainly along the Atlantic seaboard and in the northern part of the North Sea. Long-finned Pilot Whales spend most of their time offshore and, although they are widely distributed and probably occur in quite large numbers around Britain, are rarely seen until they move farther inshore during late

summer and autumn. Northern Bottlenose Whales are
probably more common than the relatively few sightings
imply and occur in deep waters off north-western Britain
mainly during the summer; they may move closer to shore
late in the season. Sperm Whales (mainly lone males or
small groups of young males) are fairly common in deep

waters beyond the edge of the continental shelf and can sometimes be seen around the Northern Isles and Outer Hebrides from late summer until early winter.

The most common baleen whale is the Minke, which is widely distributed along the Atlantic seaboard and sometimes enters the North Sea; Minkes are seen off the west coast of Scotland in spring, summer and autumn. No other baleen whales could be described as common in British waters, but several occur sporadically during the summer and are occasionally encountered (usually offshore). There are a few sightings of Humpback Whales, mainly along the Atlantic seaboard, and fairly frequent sightings of Fin Whales (which have been seen during the winter off the coast of Cornwall in recent years).

The Atlantic Frontier

There was great excitement in the mid-1990s, when Blue Whales were reported in a region known as the Atlantic Frontier by oil exploration survey ships and scientists listening in to a network of hydrophones on the seabed (originally set up by the United States Navy to track Soviet submarines travelling across the Atlantic). Lying to the west and north-west of Scotland, and stretching from coastal waters far out beyond the continental shelf in the north-east Atlantic, the Atlantic Frontier is often described as Britain's last ocean wilderness. It is widely regarded as the most important area for cetaceans in Britain, if not the whole of Europe. With a rugged and varied seabed, and depths ranging from 100m (330ft) to several thousand metres, it is believed to be an important year-round habitat for a number of species and lies on the migration routes of others. No fewer than 21 different cetaceans have been recorded altogether, including several great whales, and one particular area known as the Faroe-Shetland Channel appears to be exceptionally rich. A 17-day survey conducted by the Whale and Dolphin Conservation Society (WDCS) and Greenpeace, in the summer of 1998, recorded 303 separate sightings of an estimated 911 individual whales and dolphins of at least nine different species. The most abundant were Atlantic White-sided Dolphins and there were estimated to be 27,000 of them within the 104,087 sq km (40,178 sq miles) study area. Unfortunately, the potential of the Atlantic Frontier for whalewatching is limited by its inaccessibility: large, ocean-going vessels would be required and there are currently no commercial operators visiting the area. Oil production began in the region several years ago.

Commercial whalewatching

Closer to shore, commercial whale- and dolphinwatching did not really become established in Britain until the mid-to-late 1980s and early 1990s, but already it attracts more than 120,000 people to at least a dozen communities up and down the country. Many other places, from Duncansby Head near John O'Groats to Land's End in Cornwall, offer reasonably good to outstanding opportunities for land-based watching. A wide range of different boats are used and many operators now contribute to scientific research and monitoring through the Whale and Dolphin Conservation Society, the Sea Watch Foundation, the International Fund for Animal Welfare (IFAW) and several universities. With more than 60 per cent of Britain's coastline (much of it in undeveloped areas), Scotland undoubtedly has the greatest number and diversity of cetaceans, but Bottlenose Dolphins and Harbour Porpoises are commonly seen in England and Wales and a variety of other species turn up from time to time as well.

Information centres

There are also a number of information centres around the country, with a special focus on whales and dolphins. In Scotland, these include the Moray Firth Wildlife Centre, in Spey Bay; the Dolphins & Seals of the Moray Firth Visitor & Research Centre, in North Kessock; and the Hebridean Whale & Dolphin Trust Marine Discovery Centre, in

Moray Firth Wildlife Centre

Overlooking the mouth of the River Spey, with superb views over the Moray Firth, the Moray Firth Wildlife Centre provides a great vantage point for shore-based dolphinwatching. A new exhibition in the 18th century buildings is run in partnership with the Whale and Dolphin Conservation Society and is staffed by WDCS personnel and volunteers. It includes an audio-visual introduction to the dolphins and other wildlife of the Moray Firth and there are lots of information displays. Entry is free.

Tobermory on the Isle of Mull. In Wales, there is the Cardigan Bay Marine Wildlife Centre, in New Quay, while England has Durlston Country Park Visitor Centre, near Swanage. These centres vary enormously in their scope, but typically have information about whales, dolphins and porpoises, details on local research and conservation projects, and a shop with cetacean-related merchandise. They frequently serve as the focal point for gathering and disseminating information about whale and dolphin sightings and some offer facilities for watching from shore.

Cornwall (England)

The waters around Cornwall are home to several different species but, since it is particularly difficult to predict where and when they might appear, there are no dedicated boat trips. The best option for whalewatching is to take a ferry to the Scilly Isles, or to join an offshore birdwatching trip, both of which can be rewarded with good cetacean sightings. The western approaches to the English Channel, including the Isles of Scilly, is probably the best area in England for whale- and dolphinwatching. In addition, there are general wildlife surveys, looking for Basking Sharks, seals and seabirds as well as cetaceans, aboard an 11.7m (39ft) sloop based near Falmouth. Whales, dolphins and porpoises can also be seen on more general wildlife tours and from land-based observation points. Bottlenose Dolphins belonging to the resident (but nomadic) group that divides its time between Dorset, Devon and Cornwall can be seen sporadically. Harbour Porpoises are also a fairly common sight (especially in winter), while Risso's Dolphins, Short-beaked Common Dolphins and even Killer Whales and Long-finned Pilot Whales turn up from time to time, mainly in the extreme west. Minke Whales are the commonest baleen whales close to shore and, in recent years, small numbers of Fin Whales have been observed during the winter. The best observation points on land include Lizard Point, Gwennap Head and Cape Cornwall, but there are many others on prominent headlands.

Dorset (England)

Bottlenose Dolphins can be seen almost anywhere along the Dorset coast, although they occur only sporadically. Most of them belong to the smallest of the three known resident Bottlenose Dolphin populations in Britain and they are also seen in Devon and Cornwall. They are monitored by Durlston Marine Project, based at Durlston

Country Park just south of Swanage. The project uses a permanent hydrophone installation offshore to listen for passing animals in this part of the English Channel and the Visitor Centre has a speaker that makes it possible (with a lot of luck) to hear the animals 'live'. They are infrequently seen from shore here – the best months are April, May, October and November. More than 30 Bottlenose Dolphins have been identified in Dorset waters and Durlston Marine Project offers a popular dolphin adoption scheme.

Below: Killer Whales, or Orcas (Orcinus orca), occur in small numbers along the Atlantic seaboard of Britain and in the extreme north of Scotland.

While there are no dedicated dolphinwatching boat trips, there are trips from Swanage throughout the summer to observe the coastal wildlife and scenery and they sometimes encounter dolphins. Long-finned Pilot Whales are occasionally recorded in the area, and Short-beaked Common Dolphins and Harbour Porpoises pass by very rarely.

Pembrokeshire (Wales)

St. David's Head is the westernmost point of mainland Wales and offers an excellent vantage point for land-based porpoisewatching. It faces Ramsey Island and the waters in between are renowned for Harbour Porpoises; they are most frequently seen on the southern side, during ebb tides. Nearby Strumble Head, which lies to the north-east, is also very good – some say even better. Boat trips to the Smalls, which are about 40km (25 miles) offshore, often encounter Bottlenose, Short-beaked Common and Risso's Dolphins, and there are occasional sightings of Minke and Long-finned Pilot Whales. Killer Whales are seen on rare occasions in spring and Fin Whales in summer.

Ceredigion (Wales)

Cardigan Bay is best known for being the home of one of only three resident populations of Bottlenose Dolphins known in Britain. The town of New Quay is the place to join two-hour, four-hour or full-day boat trips to see them, at almost any time of the year. The harbour wall at nearby New Quay Head is a popular shore lookout. There are a number of other good places to see them from shore, including almost any headland from Dinas Head north through New Quay to Aberaeron. Harbour Porpoises can also be seen fairly close to shore and on boat trips, particularly during the summer and autumn. It is sometimes possible to join longer mini-expeditions, lasting up to five days, to look for whales and dolphins farther out into the Irish Sea towards Wexford. This far offshore, it is possible to find Short-beaked Common and Risso's Dolphins, Long-finned Pilot Whales, Killer Whales, Minke Whales and occasionally Fin Whales. Most of these longer trips are conducted in the summer, but a few take place later in the year, in the autumn or even during winter, when Long-finned Pilot Whales are a possibility.

Isle of Man

There is no dedicated whale- and dolphinwatching around the Isle of Man, but sharkwatching trips run by The Basking Shark Society have frequent cetacean sightings during summer. Small groups of Bottlenose Dolphins, typically comprising four to six animals, are common from May to August, although they are usually quite difficult to approach. Risso's Dolphins are also quite common and can be encountered any time from May to July; they are normally seen in groups of 20–25 individuals (although as many as 100 or more have been encountered together) and, unlike the Bottlenose, they frequently bow-ride. Short-beaked Common Dolphins are seen only occasionally, mainly from May to August.

If the sea is calm, Harbour Porpoises are relatively easy to find throughout the summer. Minke Whales are seen on shark trips about once a week (usually rising to once every couple of days during the last two weeks of July and the first week of August). Killer Whales are reported around the island most weeks from May to September (usually in pods of two to four but more than 30 have been reported

Below: *Minke Whales* (Balaenoptera acutorostrata) *are the commonest large whales around Britain and can even be seen from shore in some parts of the country, such as Cornwall and Skye.*

Above: *The beautiful island of Skye is excellent for whalewatching: it is possible to see several different species, such as Minke Whales (Balaenoptera acutorostrata) and a variety of dolphins, during boat trips and even from the shore.*

together) and Long-finned Pilot Whales are seen by fishermen and others a few times each summer.

Channel Islands

Bottlenose Dolphins are resident in the Channel Islands and are often seen around Guernsey and along the east and south coasts of Jersey. the greatest numbers occur between May and August. Minke Whales visit regularly in summer, though in small numbers.

Inner Hebrides (Scotland)

The Isle of Mull was one of the first places in Britain to offer dedicated whalewatching trips, in 1989. Sea Life

Surveys in conjunction with the Mull Cetacean Project (now a registered charity known as the Hebridean Whale and Dolphin Trust) have been taking whalewatchers and scientists around the islands of the Inner Hebrides to watch Minke Whales and other cetaceans for more than a decade. There is now a selection of whalewatching businesses to choose from. Many of the Minkes here can be individually identified, by the distinctive markings on their dorsal fins and flanks, so it is possible to learn something about particular whales that have been observed in the area for many months or years. Although Minkes are the main draw, the waters around Mull and its neighbouring islands are also home to Bottlenose, Short-beaked Common and Risso's Dolphins and Harbour Porpoises, and there are very occasional sightings of White-beaked and Atlantic White-sided Dolphins. Killer Whales and Northern Bottlenose Whales are infrequent visitors and Long-finned Pilot Whales put in an occasional appearance. The ferries between the islands, and between the Inner Hebrides and Outer Hebrides, are another good way of finding a variety of different species. One of the best-known spots for land-based whalewatching in the area is the Mull of Oa, on the southern coast of Islay. From spring 2002, the new Islay Marine Interpretation Centre (home of the Hebridean Bottlenose Dolphin Project) will be open in Lagavulin Bay.

North of Mull, the waters around the islands of Rhum and Canna, the Sound of Sleat and southern Skye have had an excellent track record for Minke Whale sightings in recent years. Harbour Porpoises are commonly seen in the area and Bottlenose Dolphins, Risso's Dolphins and Long-finned Pilot Whales are regular visitors too. Dedicated whalewatching trips lasting up to three hours, as well as weekend expeditions, are run by Sea.fari Adventures on Skye. In August 1998, two Northern Bottlenose Whales visited Broadford Bay on the Isle of Skye and stayed for some weeks; this species has been recorded annually in the region in recent years.

Outer Hebrides (Scotland)

With so many rugged headlands jutting into the north-east Atlantic, there are some excellent land-based observation points in the Outer Hebrides. Some of the better-known ones are on Lewis: Rubha Robhanais (Butt of Lewis) in the far north, Gallan Head in the mid-west, and Tiumpan Head

Below: *Humpback Whales (Megaptera novaeangliae) are by no means common around Britain, but lone individuals occasionally turn up and small numbers pass through the Shetlands each summer. They breach in a most spectacular way.*

north-east of Stornoway on the Eye Peninsula. A number of different species can be seen throughout the summer, but Risso's and White-beaked Dolphins are particularly common. Minke Whales and Harbour Porpoises are sometimes seen.

Northern Isles (Scotland)

Scapa Flow, in the Orkney Islands, made headline news twice in the 1990s when groups of Sperm Whales entered its shallow waters (six in 1993 and five in 1998). Although Sperm Whales are probably uncommon here, there have been regular sightings over the last decade, and a variety

of other cetaceans can be seen from prominent headlands such as Mull Head to the east of Kirkwall and its namesake on Papa Westray. Harbour Porpoises and White-beaked Dolphins are the most commonly seen species in summer. The Shetlands are good for land-based whalewatching and, although any prominent headland can produce sightings, two of the better-known spots are Sumburgh Head in the south and Esha Ness in the north-west. Minke Whales, White-beaked Dolphins and Harbour Porpoises are seen often, but with a little luck it is also possible to see Atlantic White-sided Dolphins, Risso's Dolphins, Killer Whales and even some of the larger baleen whales. One, two or three Humpback Whales have passed close to shore in early summer every year since the early 1990s. Good feeding areas such as Mousa Sound and around Noss and Whalsay can attract impressive gatherings of dozens of Harbour Porpoises and more than 100 have been recorded on occasion. The ferry crossings between Aberdeen or the Orkneys and the Shetland Islands, or from Mainland to outlying islands such as Whalsay, Out Skerries, Foula and Fair Isle, offer some of the most regular cetacean sightings in Britain.

Highland (Scotland)

More people watch dolphins in the Moray Firth, where the North Sea meets the Highlands in north-eastern Scotland, than anywhere else in the country. Each year, an estimated 50,000 people join boat-based trips to see the local Bottlenose Dolphins and at least another 25,000 watch them from established vantage points on shore. This is the most northerly resident population of Bottlenose Dolphins known anywhere and is one of the best-studied cetacean populations in the world, thanks to a study by researchers at Aberdeen University's Lighthouse Field Station, in Cromarty, which began in the late 1980s. Most, if not all, the dolphins are known individually and some are familiar to the many thousands of people who adopt them through a scheme operated by the Whale and Dolphin Conservation Society. Profits from the adoption project contribute to the ongoing research into these dolphins, carried out by Aberdeen University.

Scottish Natural Heritage (the government agency responsible for wildlife and habitat conservation in Scotland), the Scottish Wildlife Trust and the Whale and Dolphin Conservation Society have been working with boat operators to develop an innovative scheme for responsible dolphinwatching. Launched in 1995, and known as the Dolphin Space Programme, its aim is to ensure that everyone respects the dolphins' space. There is a code of conduct that includes tried and tested techniques for approaching the dolphins and even sets out an agreed route from which the skippers cannot deviate (except for safety reasons). It is a voluntary programme and encourages dolphinwatchers to play an active role in its implementation through an accreditation scheme. Accredited boats fly a blue flag to show that they abide by the code and have a display board showing their approved routes, so dolphinwatchers can selectively choose responsible, participating operators and then watch to make sure that they stick to the rules.

The dolphins can also be seen from shore and there are many well-known observation points. One of the most popular is Chanonry Point, but others include the Dolphins & Seals of the Moray Firth Visitor & Research Centre in North Kessock (which has a hydrophone link to listen to the dolphins 'live') and the South Sutor at Cromarty (which provides a spectacular eagle-eye view over a vast area). Fort George, Findhorn Bay, Balintore, Inverness, Nairn, Burghead, Lossiemouth, Spey Bay, Buckie, Portknockie, Cullen and Banff are also good. Although the dolphins are present year-round, summer is best for seeing them in the Inner Moray Firth; quite a few seem to feed farther offshore during winter.

Concern has been expressed about the future of the Moray Firth dolphins – the last known resident population of Bottlenose Dolphins in the North Sea. It has even been suggested that their numbers are declining by as much as 5 per cent each year, which would make the population

Chanonry Point

Few places are better for shore-based dolphin-watching than Chanonry Point, on the Black Isle, roughly midway between Cromarty and North Kessock. It is possible to stand on the pebbly beach spit that sticks out below the lighthouse and watch dolphins in the water just a few yards away.

extinct by the middle of the century. However, these gloomy predictions are by no means universally accepted and, while sightings in some areas of the inner Moray Firth do seem to have declined, sightings in other areas appear to have increased. Perhaps more importantly, relatively little is known about what may be happening further afield in other parts of the Firth. While any small, isolated population is particularly vulnerable to a range of different threats, one thing is certain: the dolphins are very popular and many people living around the Moray Firth will do everything in their power to secure a safe future for them.

Several other cetacean species occur in the Moray Firth area from time to time. Harbour Porpoises can sometimes be seen from shore at any time of the year, while Minke Whales, White-beaked Dolphins, Risso's Dolphins and even Killer Whales and Long-finned Pilot Whales sometimes appear, usually in the outer reaches farther out to sea. A lone Humpback Whale was spotted 5km (3 miles) off the coast of the outer Moray Firth in July 2001 – the first sighting of this species in the area for nearly 20 years.

On the opposite coast, facing the Minches and the Western Isles, is the little town of Gairloch. The Gairloch Marine Life Centre and Sail Gairloch Cruises are based here, offering good boat-based and land-based marine wildlife tours. There are regular sightings of Minke Whales, Short-beaked Common Dolphins, White-beaked Dolphins and Harbour Porpoises. Risso's Dolphins are in the area, although they are more common a little farther north, and there are also a few Atlantic White-sided Dolphins. Killer Whales are sometimes seen, as well, especially in late June and July. One of the best vantage points on shore is Greenstone Point, north of Loch Ewe, with wonderful views over North Minch. It is sometimes possible to see Minke Whales here from May to October and Harbour Porpoises from April to December. Other recommended watching sites are Red Point, Sron na Carra, Carn Dearg Youth Hostel and the cliffs south of Melvaig, where there are clear views of feeding grounds close to shore. Farther south, at Ardnamurchan, the most westerly point of mainland Britain, sightings of Minke Whales, four species of dolphin (Bottlenose, Short-beaked Common, Risso's and White-beaked) and Harbour Porpoises have led to the establishment of a small interpretive centre and a land-based survey project.

CANARY ISLANDS (Spain)

Main species: Short-finned Pilot Whale, Bottlenose Dolphin inshore; Sperm Whale, various beaked whales, False Killer Whale, Short-beaked Common Dolphin, Striped Dolphin, Atlantic Spotted Dolphin, Rough-toothed Dolphin, Risso's Dolphin and other species farther offshore.

Main locations: most trips depart from the west and south-west coasts of Tenerife (Puerto de Los Cristianos, Puerto Colón, Playa San Juan and Los Gigantes), as well as from La Gomera, Gran Canaria and Lanzarote.

Types of tours: half- and full-day tours, inter-island ferries, extended multi-day expeditions, research programmes, and land-based observation points.

When to go: year-round (typically 315 of 365 days of the year are suitable for whalewatching); most bad weather days occur in December and January; best season off La Gomera is March to May.

Contact details: Proyecto Ambiental Tenerife: www.interbook.net/personal/delfinc
Society for the Study of Cetaceans in the Canary Archipelago: tel/fax: +34-922-296035
Roaz Educational Whale Watching: tel: +34-922-226113
Outside the Canaries: Atlantic Whale Foundation – Canary Nature: www.whale foundation.org.uk
M.E.E.R.: www.m-e-e-r.org
Colibri Umwelt Reisen: www.colibri-berlin.de
Discover the World: www.arctic-discover.co.uk

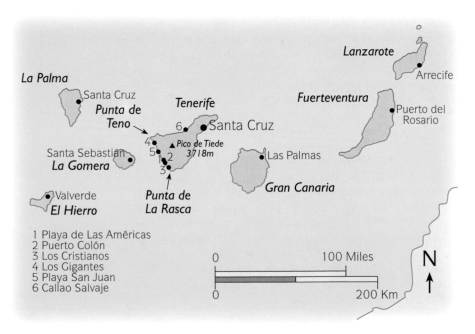

1 Playa de Las Américas
2 Puerto Colón
3 Los Cristianos
4 Los Gigantes
5 Playa San Juan
6 Callao Salvaje

In terms of sheer numbers, the Canary Islands are the most important whalewatching destination in Europe. In fact, they are one of only three places around the world that can boast more than a million whalewatchers every year (the others are the United States and Canada). Most people join short half-day trips to see resident populations of Short-finned Pilot Whales and Bottlenose Dolphins, but no fewer than 26 cetacean species have been recorded around the archipelago and there is huge potential for whalewatching farther offshore.

An autonomous community of Spain, the Canaries lie just over 100km (60 miles) off the north-west coast of Africa and some 1,100km (680 miles) south-west of mainland Spain. They consist of seven main islands (Tenerife, La Gomera, La Palma, El Hierro, Lanzarote, Fuerteventura and Gran Canaria) and several uninhabited ones. Most of the whalewatching takes place off the west and south-west coasts of the largest island in the group, Tenerife, mainly in the channel between Tenerife and neighbouring La Gomera. There is also excellent whale- and dolphin-watching from La Gomera itself, as well as from Gran Canaria and Lanzarote.

Whalewatching began in the Canaries in the late 1980s and has grown exponentially in the years since. There are now more than 50 registered whalewatch boats altogether and, on average, nearly half of them are out on the water every day. Each boat operates two to four trips daily. They range from small sailing boats and rigid-hulled inflatables to glass-bottomed boats and enormous motor cruisers, although a major proportion of them are catamarans. Two vessels have the capacity for more than 200 passengers and at least five others can carry 150 or more. Innumerable private boats join them – and the whales – every day.

Conservation concerns

Whalewatching in the Canaries is big business. A conservative estimate suggests that some 20–25 per cent of all visitors to Tenerife go on a commercial whalewatching trip and this puts the value of ticket sales alone at more than US$20 million. The industry has exploded from 40,000 whalewatchers in 1991 to more than 25 times as many today and, already, it may have exceeded its carrying capacity. There is serious concern among research and conservation groups that the whales

Below: *More than a million people join whalewatching trips in the Canary Islands every year to see the resident population of Short-finned Pilot Whales (*Globicephala macrorhynchus*).*

and dolphins are suffering as a result. In terms of the number of people watching them, and the number of hours they are being watched, the whales and dolphins of the Canaries have become some of the most intensely watched cetaceans in the world. Until quite recently, there were so many incompetent and uncaring operators that many of the animals were disturbed or even injured by speeding boats and propellers.

Regulations for whalewatching in the Canaries

Regulations were introduced in November 1996 (then revised in October 2000), as part of a determined effort to develop more responsible whalewatching, and these have had promising results. Many unscrupulous operators have now been eliminated, but there are still problems: the number of whalewatch boats is limited primarily by marina space, rather than good policy, and many operators continue to offer non-educational trips. However, the greatest concern now is the huge number of pleasure boats and jetskis, which are totally unregulated.

There are, however, some highly professional operators who have enhanced the value of their trips considerably by offering naturalist guides and contributing to local research and conservation projects. The best operators belong to the newly-formed Confederation of Whale Watching Boats of Tenerife, which is committed to ethical whalewatching, and the best advice is to support them and their outstanding efforts. There are also opportunities to join research groups to help with their work. The golden rule, when whalewatching in Tenerife, is to choose your trip carefully – not just according to price.

Meanwhile, partly as a reaction to the intense whalewatching activity, there have been moves to establish the areas frequented by Pilot Whales and Bottlenose Dolphins as a marine reserve. This could help enormously with conservation efforts, as well as providing a valuable tool to improve the quality of the whalewatching.

Geography and climate
There are two possible reasons why the channel between Tenerife and La Gomera is so good for cetaceans. Firstly, Tenerife is dominated by the volcano Mt Tiede, which rises to a height of 3,718m (12,195ft) and provides excellent shelter from the prevailing trade winds; this causes an 'island mass effect' with calm, stable waters on the leeward side of the island. The whales certainly seem to spend most of the daytime resting – either logging at the surface or swimming slowly – and are believed to feed

primarily at night when deep-living squid migrate closer to the surface. The channel therefore seems to be an important rest-and-recuperation area. Secondly, the proximity of Tenerife and La Gomera forms a submarine canyon that drops to depths of up to 3,000m (9,840ft) and provides a rich hunting ground for deep-diving whales.

The other great advantage of Tenerife is the weather. Since the Short-finned Pilot Whales and Bottlenose Dolphins are resident, it is one of the few places in the world where it is possible to whalewatch year-round. Tours typically operate more than 300 out of 365 days a year.

There are two key areas – the northern section near Los Gigantes, between Punta de Teno and Callao Salvaje, where dolphins are the main attraction; and the southern section opposite Playa de las Americas, from Callao Salvaje to Punta de La Rasca, where Pilot Whales are the key species. There is a population of at least 500 Short-finned Pilot Whales here. According to photo-identification studies, many of them are regulars and stay in the area year-round; some are temporary residents that stay for up to three months at a time; and others appear very briefly and apparently never return. Outside the Playa de las Americas region and nearby La Gomera, they are rarely encountered in large aggregations in other parts of the Canaries.

On trips leaving the west and south-west coasts of Tenerife, the whales are sometimes found within a few minutes of leaving port, although the average is typically about 20 minutes and it can take up to an hour. The frequency of sightings does not vary excessively from month to month and very few trips fail to find them.

Some of the most exciting whalewatching opportunities in the Canary Islands, however, lie in the little-visited waters farther offshore. It is possible to find Sperm Whales (a pod of no fewer than 40 were encountered recently between Lanzarote and Fuerteventura), as well as various beaked whales, False Killer Whales, Short-beaked Common Dolphins, Striped Dolphins, Atlantic Spotted Dolphins and Risso's Dolphins. Rough-toothed Dolphins can be seen almost anywhere, but the waters south of La Gomera and around El Hierro are reputed to be particularly good during the summer. La Gomera is also good for Pilot Whales (large aggregations of up to 100 have been observed and there is a mean group size of 25 year-round) and no fewer than 21 species have been recorded here altogether. An impressive variety of other species have been recorded around the archipelago at one time or another, including Blue, Fin, Bryde's, Minke and Sei Whales (Sei have been seen quite often in recent years). There is even a record of a Northern Right Whale.

Below: *Short-finned Pilot Whales (*Globicephala macrorhynchus*) live in close-knit family groups and are such social animals that they are almost never seen alone.*

Unfortunately, with Short-finned Pilot Whales and Bottlenose Dolphins so close to shore, few whalewatch operators venture more than a few kilometres from port. One inexpensive solution is to sail on the inter-island ferries, which frequently reward whalewatchers with some wonderful – and unexpected – sightings. Bear in mind, though, that the introduction of high-speed ferries in 1999 has added a new threat to the already crowded waters of the Canary Islands. The vessels make as many as 25 crossings a day at speeds of up to 40 knots and, according to a report by marine biologists from La Laguna University in Tenerife, they regularly collide with whales and dolphins and cause significant but so far unquantifiable casualties. No effort is being made to deal with the problem and any incidents of this kind should be reported to the university.

Non-profit organizations

There are also some excellent opportunities to join non-profit organizations for a week or two, as a volunteer. Several organizations provide opportunities to help with their work. These include: the Society for the Study of Cetaceans in the Canary Archipelago (SECAC), which focuses on research, conservation and education; M.E.E.R. (Mammals, Encounters, Education, Research), a German-run organization that offers practical courses in cetacean research (the main aims are to study human–cetacean interactions and to establish a marine protected area off La Gomera, designed for sustainable whalewatching). Proyecto Ambiental Tenerife and the Atlantic Whale Foundation – Canary Nature both arrange a wide range of volunteer projects such as guiding on commercial whalewatch boats and behavioural studies; their large volunteer programme (500-plus volunteers every year) is designed specifically to help whalewatch operators to develop a responsible industry. It offers free workshops for children (9–13 years) and adults in the main whalewatching harbours every July and August.

CROATIA

Main species: Bottlenose Dolphin.

Main locations: islands of Cres and Losinj.

Types of tours: extended multi-day research programmes (beware of unlicensed boats doing casual half-day and full-day dolphinwatching trips).

When to go: dolphins present year-round, but the best times are spring and autumn (when the weather is good and there are fewer boats in the area).

Contact details: Blue World: www.adp.hr

A wide variety of cetaceans have been recorded off the coast of Croatia, including Minke, Sperm, Long-finned Pilot and Northern Bottlenose Whales, Short-beaked Common and Risso's Dolphins, and Harbour Porpoises. However, the vast majority of them are merely fleeting visitors (with the notable exception of a lone Risso's Dolphin, which, at the time of writing, appears to have set up home in Rijeka Bay, to the north of Cres). The only species that can be seen predictably is the Bottlenose Dolphin.

Above: *A lone Risso's Dolphin (*Grampus griseus*) appears to have set up home in Rijeka Bay, to the north of Cres, in the Croatian Adriatic. Note the extensive scarring on its robust body as it begins to breach.*

Adriatic Dolphin Project

Small groups of people have been taken to observe a population of Bottlenose Dolphins since the early 1990s, through the Adriatic Dolphin Project. Volunteers are able to spend time with the animals and help with a conservation and research project at the same time.

The project was initially started by the Milan-based Tethys Research Institute, in 1996, and is now run by a Croatian non-governmental organization called Blue World, in collaboration with the Croatian Natural History Museum. It works specifically with the dolphins around the Cres-Losinj archipelago – the northernmost group of islands in the Croatian sector of the Adriatic Sea (a semi-enclosed basin within the Mediterranean, bordered by Italy in the west and north and by Slovenia, Croatia, Albania and Greece in the east).

The project accepts up to five paying volunteers at one time, for two-week research periods, and they cover 80 per cent of the overall costs. Everyone joins in with the day-to-day running of the project, which includes everything from meal preparation and field base maintenance to studying the dolphins at sea and logging data into a computer. Slide shows are sometimes given to local people or tourists and participants can help to organize these, or even speak if they wish.

The expeditions are shore-based, operating from the old village of Veli Losinj on the east coast of Losinj island, and the project's research vessel is used daily. It can take a few minutes or a few hours to find the dolphins, but they are seen on most days. When they are located, everyone on board has a specific task to fulfil: project participants normally record details such as behaviour and the names of individual dolphins as they are called out by the researchers. The work focuses on dolphin numbers, feeding ecology and behavioural activities, and is investigating the impact of human activities in the area. It is particularly important to monitor the status of the population, since Short-beaked Common Dolphins used to share these waters but have almost disappeared in recent years (they have been encountered only four times in the past decade). All marine mammals have been protected in Croatia since 1995, but increasing boat traffic and pollution make any dolphins in the area vulnerable.

The research team has logged 120 individually identified Bottlenose Dolphins so far and, although some of these have inevitably died, the overall population around Losinj and Cres is estimated to be about 120 animals.

Award-winning project

Since the preliminary survey was undertaken by researchers from the Tethys Research Institute, in 1987, the study of Bottlenose Dolphins off the coasts of Losinj and Cres has become the longest consistent study of Bottlenose Dolphins in the Mediterranean. It was declared one of the 'Ten Greenest Projects' by the Italian Commission Carnia Alpe Verde, in May 1995, after the researchers completed a management plan for the dolphins and succeeded in persuading the Croatian government to designate their core habitat as a marine sanctuary. Among its many other accolades, in 1999 it was awarded the Green Ribbon for Best Environmental Project in Croatia by the Croatian Ministry of Culture.

DENMARK

Main species: Harbour Porpoise.

Main locations: Kerteminde, Rømø.

Types of tours: half-day tours, research programmes, and land-based observation points.

When to go: porpoises present year-round but best conditions and most boat trips June to September.

Contact details: Fjord & Bælt: www.fjord-baelt.dk
VisitDenmark.com: www.visitdenmark.com

There are believed to be as many as 100,000 Harbour Porpoises living in the coastal waters of Denmark and, since the mid-1990s, they have been the main focus of the country's whalewatching activities. There are boat trips to see them around Kerteminde, in central Denmark, and occasional opportunities for porpoisewatching from shore on the island of Rømø, in the south-west. Although the Harbour Porpoise is considered to be the only resident cetacean here, there are at least two other species farther offshore and general summer cruises in the north-west (where the Skagerrak meets the North Sea) report regular sightings of White-beaked Dolphins and occasional Minke Whales.

There are superb opportunities to see Harbour Porpoises in the Great Belt (Store Bælt), just outside Kerteminde. It is estimated that as many as 5,000 of them live in this region year-round. The research institution Fjord & Bælt takes groups of up to 30 people on two- to three-hour porpoisewatching trips several times a week in June, August and September and every day in July. Porpoises are seen on about 90 per cent of the trips. However, sightings are very weather-dependent: if the sea is calm it is often possible to see as many as 10–15 animals, but only one or two if there is too much of a swell or chop. Even if it is too rough to find them, though, it is often possible to hear their vocalizations via the boat's on-board hydrophone.

The land-based porpoisewatching from the island of Rømø can also be excellent. The best time to watch for them is on a flood tide during the summer, and the best place tends to be from the southern end of the island, looking southwards over the water between Rømø and the German island of Sylt. There are also opportunities for porpoisewatching from the ferry that runs between Havneby, on the southern end of Rømø, and the German town of List, at the northern end of Sylt.

Captive research

Scientists from the Fjord & Bælt centre are studying captive porpoises as part of an investigation into the huge number of wild porpoises that strand or drown in fishing nets every year; the latest estimate is that 2,000–3,000 are killed annually.

THE FAROE ISLANDS (Denmark)

Main species: Long-finned Pilot Whale, Atlantic White-sided Dolphin, White-beaked Dolphin, Harbour Porpoise; potential for Northern Bottlenose Whale.

Main locations: nature boat tours depart Tórshavn, Vestmanna and several other towns and villages; good land-based watching in Vestmannasund.

Types of tours: half- and full-day tours, extended multi-day expeditions, land-based observation points.

When to go: May to September (peak period for Pilot Whales July, August and early September).

Contact details: Sp/F Skúvadal: www.puffin.fo/skuvadal
Palli Lamhauge (Vestmannabjørgini Sp/F): www.sightseeing.fo
Smyril Line: www.smyril-line.fo
Koltur Boat Trips: www.puffin.fo/koltur
Jóan Petur Clementsen: tel: +298-286119
North Atlantic Marine Activity Ltd: tel: +298-12499
Faroe Islands Tourist Board: www.tourist.fo

Outside the Faroe Islands: Arctic Experience: www.arctic-discover.co.uk

The Faroe Islands are probably better known for killing whales rather than watching them, but nature tourism here is already well established and there is considerable potential for whale- and dolphinwatching.

Located roughly halfway between Scotland and Iceland, in the north-east Atlantic, this self-governing Danish archipelago consists of 18 islands connected by bridges and small ferries. This is where the warm waters of the Gulf Stream meet the cold waters of the north, and the surrounding seas are biologically very rich. There are several tour companies offering excursions to view the spectacular seabird cliffs and seals on the rocks below. Some of them advertise the chance of encountering whales and dolphins (the vessel *Silja*, operated by Palli Lamhauge, even has a high flying bridge specifically to spot cetaceans) although, as yet, none offer dedicated trips.

A number of cetacean species can be seen around the Faroe Islands during the summer. Atlantic White-sided Dolphins, White-beaked Dolphins, Harbour Porpoises and

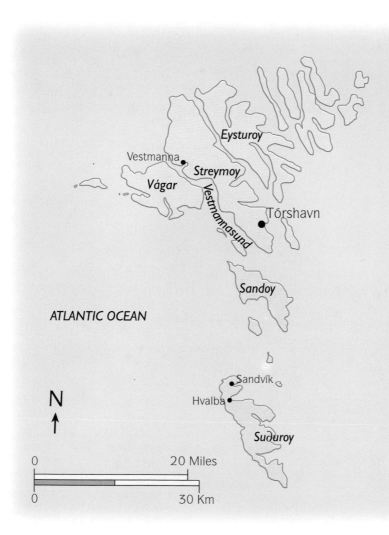

Long-finned Pilot Whales are the commonest species. They can be encountered almost anywhere – even near the capital Tórshavn – although Vestmannasund between the islands of Streymoy and Vágar has a growing reputation as something of a hotspot for land-based whalewatching. It is not unusual to see hundreds or even a thousand Pilot Whales travelling together in one enormous pod (known locally as a 'grind') on migration past the islands. A variety of other species are reported by fishermen working farther offshore, including Fin, Sei and Minke Whales, Sperm Whales and Killer Whales.

Above: *Better known for killing whales than observing them, the Faroe Islands nevertheless have considerable potential for whale- and dolphinwatching.*

Northern Bottlenose Whale

Perhaps the most exciting potential for whalewatching here is for Northern Bottlenose Whales. There are few places in the world where it is possible to see this little-known species with any regularity, but its occurrence in the Faroes may be sufficiently predictable to support a small, seasonal whalewatch industry.

There is a long tradition of hunting Northern Bottlenose Whales in the Faroe Islands and small numbers are still taken in an opportunistic drive fishery of pods sighted very close to shore. Records began in 1584 and they are unbroken from 1709 to the present day. Taking into account both stranded whales and ones that have been actively hunted, more than 800 have been recorded altogether during the past three centuries. Most of them were immature males or mature females with juveniles and most have been alone or in small pods containing up to seven animals. They have been landed at 28 different locations around the islands and washed up dead at 22 others. The two northernmost villages on Suðuroy, the southernmost island, dominate the catch – 50 per cent of all records come from Hvalba and 22 per cent from Sandvík. Local fishermen also see Bottlenose Whales quite frequently in the south, and report sightings in other parts of the archipelago as well. The peak period for all these killings, strandings and sightings appears to be 20 August to 20 September and 90 per cent occur during August, September and October.

Whaling

Long-finned Pilot Whales have also been hunted in the Faroe Islands for several hundred years. During noisy drives that frequently take many hours, entire pods are herded into sandy bays by men in small boats. More than a quarter of a million have been killed since the early 1700s, in about 1,700 different drives. The official number taken each year ranges from zero to a peak of 4,325 (recorded in 1941) and the average during the past decade has been about 1,200 whales a year. Any other whales or dolphins caught up in the drives are killed as well. The Faroese defend the hunt vigorously, arguing that it is a long-standing tradition and provides a free and welcome source of protein, but it is very strongly criticized by conservation groups as no longer necessary in such a modern society that has a relatively high standard of living.

There seems to be a genuine desire within the Faroe Islands to promote and encourage nature tourism, however, there is still a deep-rooted determination to continue hunting whales and dolphins. Consequently, many foreigners are unwilling to visit the islands on moral grounds and, until this situation changes, the potential for developing a whalewatch industry will be severely limited.

FRANCE

Main species: Fin Whale, Sperm Whale, Cuvier's Beaked Whale, Long-finned Pilot Whale, Short-beaked Common Dolphin, Striped Dolphin, Bottlenose Dolphin, Risso's Dolphin, Harbour Porpoise.

Main locations: Brittany, Normandy, Provence.

Types of tours: half- and full-day tours, extended multi-day expeditions, and land-based observation points.

When to go: May to September for most cetaceans, although many species are present year-round.

Contact details: CETUS: www.mygale.org/08/cetus
Baleines et Dauphins sans Frontières: www.bdsf.net
Hotel Kastell An Daol: tel: +33-2-9807-3911

France is home to a wide variety of whales, dolphins and porpoises and has huge potential for whalewatching. It is a relatively small industry at the moment, with fewer than 1,000 people joining commercial trips each year, but the recent designation and high profile of the Ligurian Sea Cetacean Sanctuary should attract more interest in the future.

There are two separate coastlines: one on the Mediterranean and the other, which includes the English Channel and the Bay of Biscay, on the North Atlantic.

The Mediterranean coast

The Mediterranean coast undoubtedly has the greatest potential for whalewatching, and this is where the vast majority of operators are currently based. The tours here are generally of a very high quality and there is a strong commitment to research, education and conservation. In fact, several have been developed in partnership with research projects and, during extended trips (lasting from three to eight days) participants are encouraged to help with the studies.

The species most frequently seen in the Mediterranean are: Fin Whale, Sperm Whale, Long-finned Pilot Whale, Cuvier's Beaked Whale, Risso's Dolphin, Striped Dolphin, Short-beaked Common Dolphin and Bottlenose Dolphin. Many of these can be seen year-round from car ferries that ply the waters between France and Corsica or Sardinia. There are also dedicated whale-watching trips departing from Toulon, Cannes and Nice, as well as an opportunity to swim with dolphins some 30km (19 miles) offshore.

Mediterranean Fin Whales

Mediterranean Fin Whales are genetically distinct from their contemporaries in the North Atlantic. It is thought that they belong to a geographically isolated population that lives in roughly the same area year-round. They probably make short migrations within the Mediterranean.

The most sought-after whales are the Fin Whales, which spend every summer just offshore. One company, offering trips from Cannes, virtually guarantees sightings by using a spotter plane to find the whales before directing its vessel to them. From May to September (with a peak in July).

Fin Whales gather to feed in an area called the Ligurian Sea, enclosed by northern Italy, south-eastern France and northern Sardinia. There could be as many as 900 Fin Whales here each summer, accounting for a considerable proportion of the Mediterranean population.

Below: *The Ligurian Sea, in the northern Mediterranean, is the summer home of a large population of Fin Whales (*Balaenoptera physalus*). Mediterranean Fin Whales are genetically distinct from their North Atlantic contemporaries.*

The Ligurian Sea
The Ligurian Sea has long been regarded as a critical region for cetaceans. As well as the Fin Whales, it is believed to be home to some 25,000–40,000 dolphins during the summer. In fact, there are probably between two and four times as many cetaceans here than in most other parts of the Mediterranean (the Alborán Sea also has a high abundance).

Conservation and research groups, concerned about the many threats to the region, campaigned throughout the 1990s for the sea to be given better protection. Finally, at a meeting held in Rome on 25 November 1999, the Ministers of the Environment from France, Italy and Monaco signed a treaty designating an area twice the size of Switzerland (approximately 100,000 sq km (38,600 sq miles)) as the Ligurian Sea Cetacean Sanctuary. The new sanctuary lies within an area marked by two lines drawn from Capo Falcone (north-western Sardinia) to Toulon (southern France), in the west, and from Capo Ferro (north-eastern Sardinia) to Fosso Chiarone (northern Italy), in the east.

This is the first time that several northern hemisphere countries have established a marine protected area in international waters. The treaty commits France, Italy and Monaco to coordinate research and monitoring activities in the Ligurian Sea and to intensify efforts to protect the region from pollution, destructive fishing methods and

disturbance caused by high levels of boat and ship traffic. They also have to coordinate public awareness campaigns to inform locals and tourists alike about the sanctuary and its inhabitants. Only time will tell how much impact sanctuary status will have on the cetaceans that make it their home.

The future for the whales and dolphins of the Ligurian Sea looks brighter already. Driftnets are being phased out from European waters by the European Commission, there will be a ban on offshore boating competitions and there will even be proper supervision of whalewatching activities in the area.

The French Atlantic coast

On the French Atlantic coast, in Normandy and Brittany, it is possible to join trips to see small whales and dolphins from six different communities. There are also some recognized land-based observation points at Cap de la Hague, Flamanville and Côtes des Iles. The main species encountered are: in Normandy, Long-finned Pilot Whales, Risso's Dolphins, Short-beaked Common Dolphins and Bottlenose Dolphins; and, in Brittany, Striped Dolphins, Bottlenose Dolphins and Harbour Porpoises.

Early in the summer of 2001, a Bottlenose Dolphin appeared off the west coast of France near the harbour of Port Joinville, on Ile d'Yeu, in the Bay of Biscay. It arrived with a group of other dolphins but for some reason, when they swam away, it stayed behind. The locals named the Bottlenose Moana, in memory of a Tahitian boy saved by dolphins, and it has set up home at Port Joinville. Although swimming in the harbour is banned, many adults and children have played in the water with it. Several have been bitten, however, and the local authorities have now issued safety warnings.

Preferring human company

Why dolphins should live apart from their social groups and seek out human company instead is a mystery. They rarely solicit food and, since some still mix with other dolphins, they do not seem to be social outcasts.

GERMANY

Main species: Harbour Porpoise.

Main location: Island of Sylt.

Types of tours: occasional dedicated boat trips, Sylt-Rømø ferry, in-water encounters from shore, and land-based observation points.

When to go: best May to October.

Contact details: Sylt interaktiv: www.sylt.de

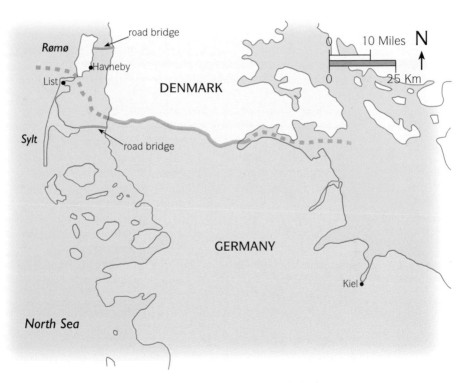

With such a small coastline facing the relatively unproductive North Sea, Germany has rather little potential for whale- and dolphinwatching. However, there are opportunities for porpoisewatching in the extreme north of the country and one operator, in particular, has been taking people to see Harbour Porpoises since the early 1990s.

There is some interest in porpoises among the 20,000 residents of the island of Sylt, just south of the border with

Denmark, and there is an ongoing study of them. There is also local support for efforts to protect the porpoise population in this part of Germany. Harbour Porpoises are under severe threat from fisheries in the North Sea (it has been estimated that some 7,500 of them drown accidentally every year in fishing nets set mainly for Cod, Turbot and Plaice) and between 50 and 100 strand on the German coastline every year.

Sylt is a popular tourist destination, attracting more than half a million people (mainly Germans and Danes) each summer, but only very small numbers of tourists have taken the opportunity to join the occasional boat-based porpoisewatching trips. Quite a few people do, however, watch the animals from shore – mainly along the beaches of Sylt and from observation points looking towards the island of Rømø, in south-western Denmark. The best time to watch for them is on a flood tide during the summer.

Also, while Harbour Porpoises are relatively nervous of people in most parts of their range, they have a reputation for being more 'friendly' in this corner of Germany and will sometimes approach swimmers in the water at Sylt. There are no organized swims, but the porpoises frequently come very close to popular beaches and many people have taken advantage over the years by joining them for in-water encounters.

Below: *The only porpoise living in Europe, the Harbour Porpoise* (Phocoena phocoena) *is now severely threatened by fisheries in the North Sea.*

There are also opportunities for porpoisewatching from the ferry that runs between List, at the northern end of Sylt, and Havneby, on the southern end of the Danish island of Rømø.

GIBRALTAR (Britain)

Main species: Short-beaked Common Dolphin, Striped Dolphin, Bottlenose Dolphin; Killer Whales farther offshore.

Main locations: Algeciras Bay: trips leave from Sheppards Marina, Queensway Quay Marina and Marina Bay Complex; also from La Línea (Spain).

Types of tours: 1½–2 hour and full-day tours, extended multi-day packages, land-based observation points.

When to go: tours year-round; Short-beaked Common Dolphins present year-round (largest numbers June to December), Striped Dolphins present year-round but mainly

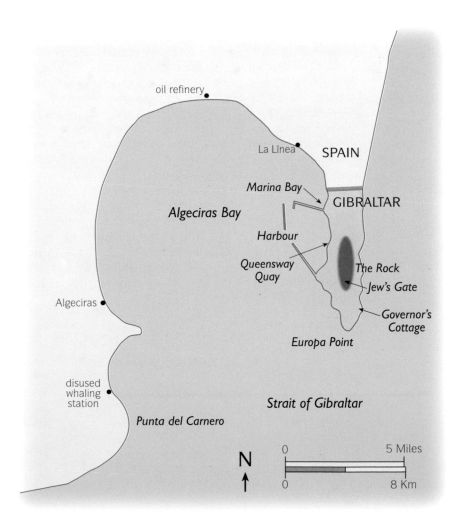

May to December, and Bottlenose Dolphins sporadic year-round but most common in summer; Killer Whales, Long-finned Pilot Whales, Sperm Whales and Fin Whales (mainly offshore but occasionally enter the Bay) in the Strait, mainly July, August and early September.

Contact details: The Original Dolphin Safari: www.dolphinsafari.gi
Fortuna: www.fortuna.gi or www.dolphinsafarispain.com
Nimo: The Helping Hand Trust: www.gibraltar.gi/helpinghand/home.html

Outside Gibraltar: Discover the World: www.arctic-discover.co.uk
WildOceans: www.wildwings.co.uk/wildoceansintro.html

Gibraltar is the birthplace of whale- and dolphinwatching in Europe. The first commercial trips took place in 1980, when Mike Lawrence began to take people out on his boat to see the local dolphins. The industry has grown substantially since then and now it is estimated that one in every six visitors to The Rock joins organized tours to encounter several different species in the Bay. The dolphins may one day rival the semi-wild Barbary Apes, or macaques, which have long been the most famous of all Gibraltar's tourists attractions.

A giant block of limestone close to the point where the Atlantic Ocean meets the Mediterranean Sea, Gibraltar forms the eastern border of Algeciras Bay. The Bay is considerably larger than Gibraltar itself, covering an area of 16 sq km (6 sq miles) compared to just 6.5 sq km (2½ sq miles) of land. It is relatively sheltered and is often flat calm, although the sea can be quite choppy or rough with a swell.

Dolphins

The dolphins are present year-round, although their numbers increase during the summer and there can be several hundred of them in the Bay until December. There is an average sightings success rate of 95–100 per cent, so they are encountered on virtually every trip. The Original Dolphin Safari normally runs 1½–2 hour trips, but will stay on the water longer if the dolphins are difficult to find.

The Short-beaked Common Dolphin is by far the commonest and about 80 may be permanent residents; many of them can be recognized by researchers and tour operators and they have been given names such as Snowy (which has a white patch on the tip on its dorsal fin) and White Spot (with a white patch at the base of its dorsal fin). They are joined by visiting dolphins throughout the year, particularly from June to September. Striped Dolphins are also fairly common and are present in the largest numbers during the summer, from May to December. Bottlenose Dolphins are more sporadic and, although they can be seen in any month, are also more frequently encountered during the summer. The Bay is believed to be used as a calving ground (although there is some evidence that Bottlenose Dolphins calve outside the Bay) and as a safe haven for young calves in their first few weeks.

Below: *Excited young dolphinwatchers have a close encounter with a Short-beaked Common Dolphin (Delphinus delphis) off the coast of Gibraltar, in Algeciras Bay.*

Land-based observation points

There are some good land-based observation points for dolphinwatching. The best ones in Gibraltar are Europa Point, which is most productive in strong easterlies when dolphins often play in the surf nearby, as well as Jews Gate and Governor's Cottage (both run by the Gibraltar Ornithological and Natural History Society and The Helping Hand Trust), but with a keen eye and a good pair of binoculars it is even possible to spot cetaceans from the top of the 427m (1400ft) Rock.

All cetaceans in the territorial waters of Gibraltar are protected by law (The Helping Hand Trust is the government adviser on its implementation).

Whales

Many other cetaceans occur regularly in the Strait of Gibraltar, although they rarely enter the Bay. Fin Whales have been encountered more frequently in recent years, especially during the summer, and both Long-finned Pilot Whales and Killer Whales are occasionally seen during dolphin trips. Sometimes, Minke Whales are also seen from late May to July. In the past, very lucky dolphinwatchers have been rewarded with sightings of a range of other species, from Sperm Whales to False Killer Whales.

The larger species are more frequently seen on longer trips, which explore farther offshore. Most of these leave from nearby Spanish ports. Fin Whales pass through in small numbers throughout the year, Sperm Whales are seen mainly south-west of Tarifa from May to July, and Long-finned Pilot Whales are resident. Killer Whales spend a couple of months in the Strait each summer to feed on tuna; they are most often seen in July, August and early September. Cuvier's Beaked Whales and Risso's Dolphins are also seen from time to time and several other species have been recorded over the years.

Note: Orginally based in Gibraltar, Fortuna recently moved its operation to La Línea, just across the border, in Spain (out of frustration at the Government's failure to adopt a code of conduct for dolphinwatching in the Bay).

GREECE

Main species: Sperm Whale, Cuvier's Beaked Whale, Short-beaked Common Dolphin, Striped Dolphin, Bottlenose Dolphin, Risso's Dolphin.

Main locations: Crete, Kalamos, Korinthiakos Gulf (Gulf of Corinth).

Types of tours: half-day tours, extended multi-day research expeditions, cruise ships, shore-based observation points, inter-island ferries.

When to go: whales and dolphins present year-round, but best conditions June to September.

Contact details: Pelagos Cetacean Research Institute: www.pelagosinstitute.gr
Tethys Research Institute: www.tethys.org

It was Aristotle, the Greek scientist and philosopher, who first observed and admired dolphins off the coast of Greece. More than 2,300 years ago, he recognized that dolphins are mammals and not fish and learned to tell one individual from another by the nicks and scratches on their dorsal fins.

Casual dolphinwatching began here in the late 1980s, but it was nearly a decade before dedicated trips were properly established. These days, there are a number of commercial dolphinwatching trips, operating mainly from the Greek islands in the Ionian and Aegean Seas. Most of them last for three or four hours and they have considerable success in finding Striped, Short-beaked Common and Bottlenose Dolphins, with occasional sightings of Risso's Dolphins. There are some particularly rich spotting grounds between Corfu and Cephalonia, in the Ionian Sea, and off south-western Crete.

The waters off southern Greece are extremely deep – up to 5.5km (3.4 miles) in places – and they attract deep-water squid that are the preferred prey of Sperm Whales. Some of the dolphin tours explore areas with large numbers of Sperm Whales, but the best way to find these deep-diving cetaceans is by listening for their echolocation clicks underwater and, since the tourist boats are not equipped with hydrophones, they are rarely encountered except on dedicated trips.

Environmental groups and research

Environmental groups in Greece are concerned about the low quality of some of these half-day tours. Few have experienced naturalists on board, or provide any kind of informative commentary, and even fewer participate in research projects. There is also concern that some of the skippers are inexperienced in manoeuvring their boats around dolphins, or behave irresponsibly in the company of dolphins, and cause unnecessary disturbance. Therefore, until the standards improve, they recommend that whalewatchers join multi-day research expeditions instead.

The Tethys Research Institute runs the Ionian Dolphin Project, in the eastern Ionian Sea, and partly funds the study by taking volunteers who pay to help. The research began in 1993, specifically to investigate the social ecology of Bottlenose and Short-beaked Common Dolphins, and volunteers assist by recording and identifying individuals from small inflatable dinghies with fibreglass keels, or from three different observation points on shore. It can take anything from a few minutes to several hours to find the animals, although one hour is typical from the boats. Once they have been sighted, observations last for as long as possible. The project is based in an old Greek house in the tiny village of Episkopi, on the island of Kalamos, and no scientific background is required to participate.

*Below: Large numbers of Sperm Whales (*Physeter macrocephalus*) feed on squid in very deep water off the south-western coast of Crete. Note the distinctively-shaped flukes.*

Short-beaked Common Dolphins

Short-beaked Common Dolphins were once widely distributed in the Mediterranean Sea but their numbers have declined dramatically in recent years, and eastern Ionian coastal waters are among the few places left where they can be encountered on a regular basis. This relic population is in need of strong conservation measures and one of the long-term aims of the Ionian Dolphin Project is to study their interactions with fisheries in order to improve relations with the local fishermen.

Tethys also offers intensive eight-day workshops, designed for students and researchers wishing to gain more practical, hands-on experience in dolphin research. Held at the Ionian Dolphin Project field station, on Kalamos, they include daily seminars as well as practical field activities and cover a wide range of research methods from photo-identification to remote biopsy sampling.

The Pelagos Cetacean Research Institute has two eco-volunteer programmes: one studying Sperm Whales off south-western Crete, the other studying dolphin populations in the Korinthiakos Gulf.

The Cretan Sperm Whale Project began in 1995, when incidental reports suggested the presence of large numbers of Sperm Whales off the south-western coast of Crete. Four years later, the research team began to take paying volunteers to help fund their work. The study area consists of very steep underwater cliffs, dropping to extraordinary depths close to shore, and it is likely that this is the most important area in the Mediterranean for Sperm Whales. This particular population is unusual because maternity groups of females and their calves coexist with individual males, so they feed, calve and probably mate in the same area year-round. They are believed to be isolated from the Atlantic Sperm Whale population, which would make them particularly vulnerable to environmental changes, and the degree of genetic isolation is one of the key areas of research. Now there is a sense of urgency to the work, because there are signs that the population may be decreasing, so the ultimate aim is to provide the necessary scientific grounding for conservation guidelines to secure the future of the whales. The study is also compiling a catalogue of individuals, investigating migration routes, and studying communication patterns. Volunteers join week-long trips and help to locate the whales with binoculars, or by using the onboard hydrophone, and assist in recording observations. Sightings records suggest that Sperm Whales are in the study area (which has recently been extended to include parts of the Aegean Sea around Chalkidiki and Northern Sporades) at least 75 per cent of the time.

The Korinthiakos Gulf Dolphin Research Project also boosts funds by taking volunteers. The project began in 1995 and studies the four local dolphin species: Striped, Short-beaked Common, Bottlenose and Risso's. There are more dolphin sightings in the Gulf than anywhere else in Greece and it is widely considered to be one of the best places for dolphin encounters in the whole of the Mediterranean. Interestingly, three of the four species (Striped, Short-beaked Common and Risso's) have been observed in mixed-species schools and this is a particular focus of the study. Volunteers stay at the research base in the village of Lykoporia, on the coast of the Korinthiakos Gulf, and search for dolphins from a small inflatable boat.

There are also whale- and dolphinwatching opportunities from the numerous inter-island ferries, cruise ships and sailing charters that continually ply Greek waters.

Whales in the headlines

Pelagos is also conducting a study of Cuvier's Beaked Whales in the waters of the Aegean, Ionian and Libyan Seas. These whales hit the headlines in 1996, when a mass stranding in the Kyparissiakos Gulf, involving the deaths of at least 13 animals, was linked to military sonar tests being conducted by NATO. The tests have since stopped, albeit temporarily, and Pelagos researchers are trying to learn more about these little-studied whales and their ecological requirements. This particular research project is not open to volunteers, although beaked whales are occasionally encountered on other trips.

GREENLAND (Denmark)

Main species: Minke Whale, Fin Whale, Humpback Whale, Narwhal, Beluga, Killer Whale, Harbour Porpoise.

Main locations: Ammassalik, Disko Bay area, Qaanaaq (Thule), south-west region.

Types of tours: half- and full-day tours, kayaking, cruises and extended multi-day expeditions, research programmes, and land-based observation points.

When to go: whales present and accessible June to October (depending on species), but weather more likely to be poor September to October.

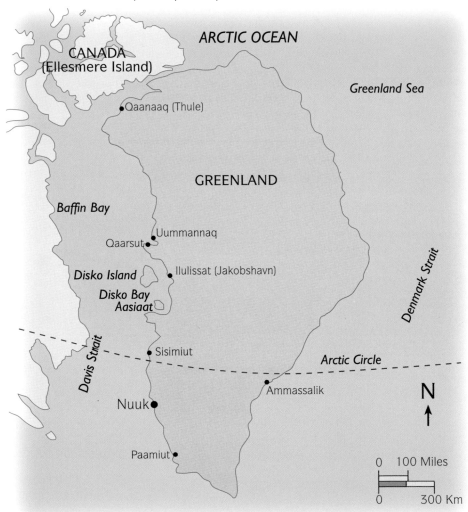

Contact details: Team Arctic: www.greenland-guide.dk/teamarctic/default.htm
77° North: www.77north.com
Greenland Tourism: www.greenland-guide.dk/gt/default.htm

Outside Greenland: Quark Expeditions: www.quarkexpeditions.com
Arcturus Expeditions: www.arcturusexpeditions.co.uk
Arctic Experience: www.arctic-experience.co.uk

A self-governing territory of Denmark, Greenland is the second largest island in the world after Australia. It has a fabulously rugged, indented coastline and such productive seas that it attracts rich populations of invertebrates, fish, seals and whales. The scenery is stunning and, during the height of summer, much of the country is bathed in the golden light of the midnight sun.

Whalewatching here is unlike anywhere else in Europe. There are superb opportunities for observing Arctic species rarely found elsewhere in the region and the spectacular icy setting adds a new dimension to watching more familiar species. Imagine watching two male Narwhals jousting with their tusks one day and a Humpback Whale breaching in front of a brilliant snowy-white iceberg the next. The local specialities are Narwhal and Beluga, but there are also good numbers of Humpback, Fin and Minke Whales, and Killer Whales are frequently seen in some areas. There are even occasional sightings of Sperm Whales and Long-finned Pilot Whales, while Blue and Bowhead Whales are undeniably rare but nevertheless possible.

Whales are still being hunted in Greenland (Minke and Fin Whales under the International Whaling Commission's aboriginal subsistence whaling regime, as well as Narwhals and Belugas) but there has been whalewatching here since the early 1990s. Dedicated tours are available from three different communities and it is possible to organize special charters from at least three others. A variety of natural history cruises, usually aboard icebreakers or ships with ice-strengthened hulls, explore long stretches of the Greenland coast and include regular sightings of whales as an important part of their itineraries. Several thousand people visit to watch Greenland's whales in a typical year, and there would probably be many more if it were not so expensive and if the whalewatching was not quite so dependent on the unpredictable weather.

Qaanaaq (Thule)

The remote north-western region is good for the high-Arctic species. The only place to stay is Qaanaaq, or Thule, which is the northernmost naturally inhabited community in the world. The local Inuit hunt Narwhal, which are fairly common in the area and can be seen most readily during the last two weeks of July and throughout August. There are no organized whalewatch tours, as such, but kayaks can be rented to get close to the whales and extended kayaking trips are sometimes arranged in conjunction with the hunters themselves. This area is also known to be a wintering ground for the endangered Davis Strait stock of Bowhead Whales.

Disko Bay area

Disko Bay and the area immediately to the north, around Uummannaq, is perhaps the best-known whalewatching region in Greenland. Whales are literally guaranteed on half-day whale safaris in Uummannaq Fjord, which usually cruise from Uummannaq to the little settlement of Qaarsut. The most commonly seen whales are Fin, which seem to be unusually curious in this part of the world and will often investigate and associate with boats, but Narwhal, Beluga and Minke Whales are all encountered on a regular basis. Occasional sightings of Humpbacks, Sperm Whales and Killer Whales are occasionally recorded. There is a hydrophone on board, so it is possible to listen to the whales as well as watch them, and the best time to go is mid-July to late October.

The town of Ilulissat (Jakobshavn) is Greenland's third largest community and, lying near the mouth of the ice fjord and overlooking Disko Bay, is surrounded by magnificent views. In summer time, this is a good area to see Fin Whales, as well as Minkes and Humpbacks, and is renowned for its Belugas, which spend October and November in the area. Narwhals are sometimes seen here, as well, although they tend to be a little farther offshore – segregated into all-male or nursery groups. The island of Aasiaat, at the southern end of Disko Bay, is also good for whales. There are boats offering trips to see Fin, Humpback and Minke Whales, as well as Belugas, surprisingly close to town, and it is possible to find Narwhals farther offshore. The best time to go for Belugas and Narwhals is September and October and, although the days are short and the weather especially

Left: *Dubbed 'sea canaries' by ancient mariners, because of their great repertoire of trills, moos, clicks, squeaks and twitters, Belugas (Delphinapterus leucas) can be seen in many areas along the west coast of Greenland.*

unpredictable at this time of year, it is worth the time and effort. Kayaks can be rented at both Ilulissat and Aasiaat and, with luck, it is sometimes possible to paddle among the whales. There are also whale- and sealwatching trips from the community of Sisimiut, roughly midway between Ilulissat and the capital of Nuuk (Godthab) farther south.

Nuuk

There are half-day trips to see Humpback Whales from Nuuk itself, down in the south-western corner of the country, and these operate throughout the summer. There is excellent whalewatching from Paamiut (Frederikshab), both on half-day or full-day trips and on extended week-long whale safaris exploring farther along the coast, and this is a popular place to photograph fluking or breaching Humpback Whales against spectacular backdrops of icebergs and glaciers. Fin and Minke Whales are also encountered in this relatively warm corner of the country, in fairly good numbers.

East coast

The east coast of Greenland has whales, too, and there have been more sightings in recent years (including more than 250 Humpbacks late in the summer of 2001). There are relatively few communities on this side of the island, but charter trips leaving from the town of Ammassalik often encounter Belugas, Narwhals and various baleen whales during the summer. Sea kayaking tours through the fjords in this region can also offer very good opportunities for sighting whales.

Shore-based whalewatching

Shore-based whalewatching is not yet well established in Greenland, but there is considerable potential. With a little luck and patience (and, of course, the right ice and weather conditions) whales of several different species can be seen from a number of coastal communities, including Qaanaaq, Uummannaq, Aasiaat, Ammassalik and even Nuuk.

ICELAND

Main species: Minke Whale, Sei Whale, Fin Whale, Blue Whale, Humpback Whale, Sperm Whale, Long-finned Pilot Whale, Killer Whale, Atlantic White-sided Dolphin, White-beaked Dolphin, Harbour Porpoise.

Main locations: Húsavík, Hauganes, Dalvík, Stykkishólmur, Olafsvík, Arnarstapi, Reykjavík, Hafnarfjörður, Keflavík, Sandgerði, Grindavík, Westmann Islands.

Types of tours: half- and full-day tours, extended multi-day expeditions, research programmes, and land-based observation points.

When to go: most tours operate from mid-April to September, although some operate from early April to the end of October and the best period for weather and whales is June to August; typically, more Humpback Whales in early summer, more Blue Whales at the height of summer and more Killer Whales in early and late summer; 24-hour daylight in late June and early July for midnight whalewatching.

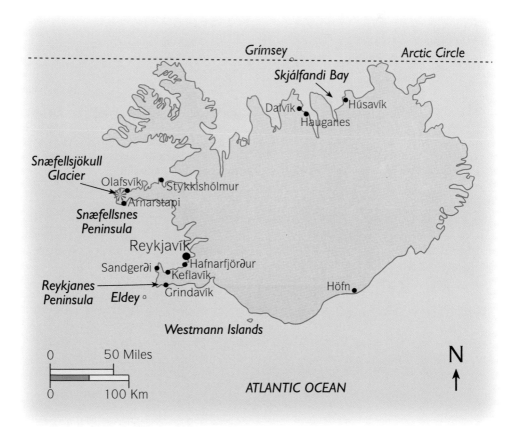

Contact details: *Húsavík:* The Húsavík Whale Centre: icewhale@centrum.is
Norðursigling (North Sailing): www.nordursigling.is or www.northsailing.is
Hauganes: Niels Jónsson: tel: +354-466-1690; fax: +354-466-1690
Dalvík: Sjóferðir (Seatours): tel: +354-466-3355; fax: +354-466-1661
Stykkishólmur / Olafsvík: Sæferðir (Sea Tours ehf): www.saeferdir.is
Arnarstapi: Sea Tours / Sölvi Konráðsson: tel: +354-435-6738; fax: +354-435-6795
Keflavík: The Dolphin and Whale Spotting company (HI Ferðapjónusta): www.dolphin.is (also operates from Grindavík and Sandgerði depending on weather and whales)
Grófin: www.marine-marvels.com
Wild Whale Watching: tel: +354-422-7292 or +354-692-8779
Hafnarfjörður: Elding Whale Watching: www.elding.is
Húni II Whale Watching: www.islandia.is/huni or www.randburg.com/is/huni.html
Reykjavík: Hvalstoðin – Whale Watching Centre: www.whalewatching.is
Westmann Islands: Viking Boat Tours: www.boattours.is

Outside Iceland: Discover the World: www.arctic-discover.co.uk

Whale and dolphinwatching in Iceland? With the country's long history of whaling, not many years ago it would have seemed about as likely as scuba-diving in Switzerland, downhill skiing in Holland, or beach holidays in Spitzbergen. However, since the first commercial trip left Höfn in 1991, aboard a 135-tonne lobster fishing boat, to see Minke Whales and Humpback Whales off the spectacular south-east coast, Iceland has become a mecca for whalewatchers from all over the world.

Land of fire and ice

Lying just below the Arctic Circle, covered in lava flows, with accessible glaciers and enormous seabird colonies, huge waterfalls and dramatic geothermal areas, and more uninterrupted elbow room than any other country in Europe, Iceland is a naturalist's paradise. It is this spectacular setting, as much as anything else, that draws whalewatchers here in droves. Where else can you encounter Minke Whales in the orange glow of the midnight sun, watch Killer Whales against a striking backdrop of snow-capped mountains, sail alongside Blue Whales less than an hour after passing one of the most famous volcanoes in the world *and* watch Humpback Whales feeding in the midst of hundreds of diving gulls and terns? With a little luck, the 'land of fire and ice' offers all this and much more besides.

*Opposite: A beautifully restored oak fishing boat takes a group of whalewatchers into Skjálfandi Bay, off the north coast of Iceland, hoping to see Minke Whales (*Balaenoptera acutorostrata*).*

Whalewatching

Below: *Humpback Whales (*Megaptera novaeangliae*), and several other baleen whale species, gather in the rich waters around Iceland every summer to feed.*

Whalewatching may be relatively new to Iceland, but there are already about a dozen whalewatch operators working from almost as many different towns and villages along the north, south and west coasts. Beautifully restored oak fishing boats, motor cruisers, large and comfortable catamarans and even vessels that were once used to hunt Minke Whales have all been customized for the purpose. Between them, they take tens of thousands of people to see many different cetaceans every year – in many cases, with an astonishing sightings success rate of nearly 100 per cent.

Many Icelandic whalewatch operators provide excursions of a very high standard. The skippers are experienced, after spending many years at sea, and are well versed in whale etiquette and behave responsibly around the animals. The on-board naturalists, who accompany all the best trips, are well-informed and provide interesting and entertaining commentaries. Like the skippers, they are highly skilled in finding – and identifying – whales, dolphins and porpoises at sea. Some operators are even beginning to use the whalewatch boats as platforms for research and they all keep detailed logbooks. There are no formal regulations for whalewatching, but all the more responsible operators are

Fast-growing industry

Iceland now has the fastest-growing whalewatch industry in Europe. According to information compiled by Erich Hoyt and the founding director of The Húsavík Whale Centre, Asbjorn Bjorgvinsson, since 1994 whalewatch numbers here have shown a phenomenal 251 per cent increase, on average, every year. It is the second highest rate of increase anywhere in the world since the mid-1990s. The industry continues to grow at an ever-astonishing rate and, in 2001, no fewer than 60,550 people joined commercial whalewatching trips from nine different locations around the country.

working to voluntary guidelines and there is a proposal to establish an official Whalewatching Organization of Iceland to encourage cooperation and professionalism.

Productive waters

The surrounding waters are highly productive and provide a rich feeding ground for whales, dolphins and porpoises. Iceland itself sits on the junction of several submarine ridges, including the Mid-Atlantic Ridge (which effectively divides the Atlantic Ocean into two distinct basins). It also lies at the point where the warm Gulf Stream from the south meets the cold East Greenland and East Icelandic Currents from the north. The resulting underwater turmoil causes a tremendous mixing of the seas and helps to carry cold, nutrient-rich water from the ocean depths up towards the surface. These nutrients encourage the growth of millions of tonnes of minute aquatic plants, or phytoplankton, which form the basis of the marine food chain. The phytoplankton provide rich pickings for minute aquatic animals, or zooplankton, and these, in turn, attract hungry fish, as well as whales and other predators.

The sheer variety of cetaceans seen around the country is quite remarkable. Roughly one quarter of all known species have been recorded in Icelandic waters and, with a little luck and good weather, whalewatchers frequently encounter three or four of them on a typical half-day trip. Anyone spending a whole week on the water, in different parts of the country, could be rewarded with nine or ten different species and, of course, may be even luckier and encounter some of the others that turn up less frequently as well.

Minke Whales, White-beaked Dolphins and Harbour Porpoises are the most common and can be seen on almost all whalewatch trips around Iceland's 6,000km (3,725 mile) coastline. Some of the Minkes are incredibly inquisitive and friendly and, indeed, many of the best photographs ever taken of this species have been captured in Icelandic waters. Blue Whales, Fin Whales, Humpback Whales and Killer Whales are also encountered frequently and, although they can be seen almost anywhere, are best seen in certain well-known hotspots. Sei Whales, Long-finned Pilot Whales and Atlantic White-sided Dolphins are more sporadic (some years tend to be better than others) and Sperm Whales can be found in deep water farther offshore. One of the great pleasures of whalewatching in Iceland, however, is that almost anything can turn up – including Northern Bottlenose Whales and other beaked whales on rare occasions. In the summer of 1997, for example, three Northern Bottlenose Whales delighted whalewatchers in Skjálfandi Bay, on the north-east coast. Normally found in deep water far out to sea, these strange-looking whales, with huge, bulbous foreheads, spent many weeks surprisingly close to shore and within easy reach of whale-watch operators based in the fishing village of Húsavík.

Húsavík

Approximately half of all Iceland's whalewatchers join trips from Húsavík. In recent years, the village has rapidly developed into one of the premier whalewatching locations in Europe and, now that it is a major tourist centre, it is positively thriving. It even has an annual Whale Watch Festival around Midsummer's Day, which attracts people from all over Europe and North America.

There are two main attractions in Húsavík: friendly Minke Whales and midnight whalewatching. The village nestles on the shore of Skjálfandi Bay and some of the Minke Whales that spend every summer here are particularly friendly and readily approach whalewatch boats. In fact, they are so friendly that it is often hard to tell who is supposed to be watching whom. On one memorable occasion, there were no fewer than 15 of these 'friendlies' around a whalewatch boat (some almost within touching distance), spyhopping, swimming underneath and lying alongside. Another once-in-a-lifetime experience is watching the Minkes in the soft, golden glow of the midnight sun – their dark bodies send ripples across the glassy surface of the sea and their blows produce colourful displays of droplets in the rays of the sun. There is also a population of Harbour Porpoises in Skjálfandi Bay and White-beaked Dolphins are regular visitors, too (many young calves are seen in mid-summer). A wide variety of other species turn up from time to time, staying for a few hours, days or even weeks, including Blue, Fin and Sei Whales, Humpback Whales and Killer Whales; Blue Whales, in particular, are becoming increasingly regular visitors in this part of the country. It is not unusual to find one species or another within an hour of leaving dock and there are several excellent vantage points in and around Húsavík for land-based watching.

Húsavík is also the home of the award-winning Whale Museum. The first of its kind in Iceland, this non-profit information centre was opened in the summer of 1998. It is a highly imaginative and informative museum, with interpretive displays on the cetacean species around Iceland, several whale skeletons, artefacts from the whaling days, and a library and video collection. It also provides a forum for coordinating the whalewatching industry and encourages more cooperation and professionalism.

Grímsey
The idyllic island of Grímsey, most famous for being dissected by the Arctic Circle, lies just 40km (25 miles) off Iceland's north coast. Whalewatch operators in Húsavík occasionally organize long day-trips here – the only place in Iceland that can claim to be truly Arctic – looking for whales on the way to and from the island, with a few hours on shore to enjoy the rich bird life or to explore the small fishing and farming community.

Snæfellsnes Peninsula
Another whalewatching hotspot is off the west coast. Half- and full-day tours leaving the village of Olafsvík sail along the rugged Snæfellsnes Peninsula, past the famous Snæfellsjökull Glacier (location for *Journey to the Centre of the Earth*), and out into the open sea. It is not unusual to see family pods of Killer Whales along the way, as well as Minke Whales, White-beaked Dolphins and Harbour Porpoises, but most of the trips head for an

area some 12–24km (7½-15 miles) offshore. Here, in the shadow of the cone-shaped 1,446m (4,745ft) volcano, is an important feeding ground for Blue Whales. Iceland is the only place in Europe – and one of just a handful of places anywhere in the world – where it is possible to find Blues with any regularity. They are seen on almost every trip (the peak time is July and August) and some years have an amazing 100 per cent sightings record. It is not unusual to see several individuals together and some trips encounter groups of ten or more. Humpback Whales and Fin Whales can also be seen feeding in the same area.

Below: *Surrounded by hundreds of Arctic Terns and other seabirds, a Fin Whale (*Balaenoptera physalus*) feeds alongside a group of whalewatchers in Iceland.*

Reykjanes Peninsula

Blue Whales are sometimes seen off the south coast, as well, although they tend to be a little farther offshore. Some of the best whalewatching here takes place around the islet of Eldey. Lying a little over 14km (9 miles) south-west of the Reykjanes Peninsula, in the extreme south-west of the country, Eldey rises spectacularly from the sea to a height of nearly 80m (260ft) and has precipitous cliffs on all sides. It is the home of a spectacular Gannet colony and the birds are literally packed shoulder-to-shoulder on the 100 x 80m (330 x 260ft) islet top. The world's last breeding pair of Great Auks also lived on Eldey – they were clubbed to death by a hunting party in 1844. When there are whales alongside the boat, and hundreds or even thousands of Gannets filling the skies above, whalewatching in this little corner of Iceland can be a very special experience. Trips leave Reykjavík, Hafnarfjörður, Keflavík, Sandgerði and Grindavík and many of them include a visit to the Nature Centre in Sandgerði as part of the tour. In recent years, they have had a sightings success rate of about 90 per cent. Humpback and Minke Whales, Killer Whales, White-beaked Dolphins and Harbour Porpoises are all seen in the

Below: *Strikingly large and robust, the White-beaked Dolphin (*Lagenorhynchus albirostris*) is the most familiar member of the dolphin family in Icelandic waters. Note its complex white, grey and black markings as it porpoises.*

area regularly. Trips from Hafnarfjörður have also been enjoying some outstanding sightings in recent years. The south-west may yet rival Stykkishólmur and Húsavík in the whalewatching stakes.

The whaling debate

The growth of the whalewatching industry in Iceland has indirectly provided a strong argument against a resumption of whaling, which is a never-ending subject of heated debate. Iceland has not officially killed a whale since the late 1980s, but a few ex-whalers (with support from a number of politicians) are determined to kill more.

Modern commercial whaling began in Iceland in 1883, when the Norwegians established the first successful whaling station at Álftafjörður, on the west coast. By the turn of the century, seven land stations were dotted along the west coast and whaling had become big business. Whale numbers soon declined, however, and most of the stations were forced to move to the east. When whale numbers declined there as well, the Icelandic Parliament banned all whaling, in 1915, to allow the stocks to recover. Some 8,000 Fin Whales, 6,000 Blue Whales and 3,000 Humpback Whales had already been killed, as well as smaller numbers of Sei and Sperm Whales.

Then, in the late 1920s, the appearance of floating factory ships in Icelandic waters, from Norway and other countries, prompted the Icelanders to re-establish their own land-based whaling industry. During the five-year period leading up to the Second World War, three vessels operating from a single station in Tálknafjörður, western Iceland, killed a total of 469 whales. After the war, in 1948, a single station on the northern shore of Hvalfjörður, in south-west Iceland, resumed whaling and continued until the end of 1985 – when all commercial whaling was banned by a worldwide moratorium under the International Whaling Commission (IWC). During this 37-year period, the total catch was 14,516 whales, with an annual average of 234 Fin Whales, 68 Sei Whales and 82 Sperm Whales (until the IWC set zero catch quotas for Sperm whaling in 1982). The catch also included five Humpback Whales (protected in the North Atlantic in 1956) and 163 Blue Whales (protected in the North Atlantic in 1961). Meanwhile, Minke Whales were also being hunted by fishermen operating small whaling vessels in Iceland's coastal waters, mainly in the north and west, with an average catch of 50 per year (1914–1950s) and then 200 per year (1950s–1985).

Despite the worldwide moratorium, Iceland continued to take a limited number of Fin and Sei Whales every year from 1986-89 'for the purposes of scientific research'. This was highly controversial, however, and, after four stormy years, was finally stopped. The argument for a resumption of whaling has been raging ever since. The main case being given is that the whales are having a negative impact on Iceland's fish stocks. Scientists at the Marine Research Institute claim that the Minke Whale population in this part of the North Atlantic consumes over 1 million tonnes of small fish and krill every year – and argue that 250 Minke Whales and 100 Fin Whales should be taken annually to limit the competition with human fisheries. At the same time, though, they admit that this would actually make little impact on the marine ecosystem around Iceland.

Revenue from whalewatching

Meanwhile, whalewatching continues to grow apace and, in the past seven years, has actually grown faster than any other sector of the Icelandic tourist industry. It now provides a significant revenue for many local communities and for the economy as a whole. Asbjorn Bjorgvinsson estimates that the direct value of the whalewatch industry in Iceland is now close to US$8.5 million. That is approaching the value of the whaling industry at its peak in the period 1950–80 (during the 'scientific' whaling years, in the late 1980s, the annual value of whaling was estimated to be a lower US$3–4 million). The other great advantage of an industry that watches whales, instead of killing them, is that it is in tune with efforts to promote Iceland as a 'natural' destination.

Resuming whaling

If whaling is ever resumed, it could have a very negative impact on whalewatching, as well as on the whales. Recent evidence suggests that many keen whalewatchers would refuse to visit a country that threatens to kill the very animals they have come to see and, ironically, it is the friendly, trusting Minke Whales of Skjálfandi Bay that would probably be the first to go.

REPUBLIC OF IRELAND AND NORTHERN IRELAND

Main species: Minke Whale, Long-finned Pilot Whale, Short-beaked Common Dolphin, Bottlenose Dolphin, Risso's Dolphin, Harbour Porpoise.

Main locations: Shannon Estuary, Dingle, Cape Clear, Northern Ireland.

Types of tours: half- and full-day tours, extended multi-day expeditions, research programmes, walking tours, and land-based observation points.

When to go: best May to October for most species and locations; several Dingle operators offer trips to see Fungie year-round; Sei and Fin Whales possible in autumn and winter; Killer Whales most frequent August to September.

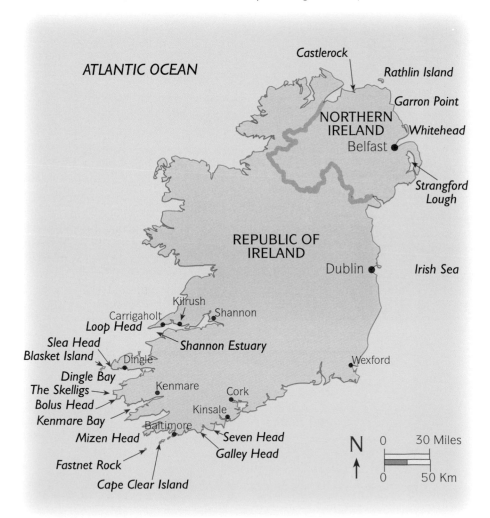

Contact details: Dolphinwatch Carrigaholt: www.dolphinwatch.ie
Dingle Boatmen's Association: www.dinglemarine.com/dtrips.html
Saoirse Sea Sports Dolphin-watching Under Sail: www.gannetsway.com
Fastnet Charters: tel: +353-28-36450
Dolphin Discovery Kilrush: www.shannondolphins.ie/sdwf/tours.htm
Shearwater Cruises: www.kinsale.ie/shearwater.htm or johnpetch@eircom.net
Rossbrin Yacht Charters: tel: +353-28-37165
Seafari Cruises: www.seafariireland.com
Irish Whale and Dolphin Group: http://iwdg.ucc.ie or www.iwdg.ie
Shannon Dolphin and Wildlife Foundation: www.shannondolphins.ie

Whalewatching in Ireland began rather suddenly and unexpectedly in 1984, when a solitary Bottlenose Dolphin arrived off the coast of Dingle, County Kerry, and decided to stay. He was adopted by the local residents (who named him Fungie) and quickly became one of Ireland's top tourist attractions. Known around the world as the 'Dingle dolphin', he has entertained more than a million visitors over the years and now attracts a phenomenal 200,000 people annually – accounting for the vast majority of whale- and dolphinwatching in Ireland. He can be seen from shore, leaping and spyhopping, and a dozen boat owners offer Fungie-watching trips (through the Dingle Boatmen's Association) at the height of the summer. They leave Dingle Pier all day, every day. There is also a two-hour early morning swimming trip, when it is often possible to get in the water with Fungie in Dingle Harbour. A multi-million pound industry now depends on his continuing presence in the area, which is quite a thought when you consider that he has already stayed much longer than most other solitary dolphins.

There are many other opportunities for watching whales and dolphins in Ireland, and in 1991 the entire region was given special protection under the Irish Whale and Dolphin Sanctuary.

Freedom of the Shannon
The Shannon Estuary, on the south-west coast immediately north of Dingle Bay, is the second biggest draw after Fungie. Commercial dolphinwatching began here in 1992 and already it attracts as many as 15,000 people each year. According to photo-identification studies carried out by University College, Cork, the estuary is home to a resident population of about 120 Bottlenose Dolphins. It is the only

known resident population of this species in Ireland. The region is considered so important that, in 1999, it was declared an EU Special Area of Conservation. There is an established management plan to secure the future of the dolphins and an accreditation scheme for tour operators (known as the 'Saoirse na Sionna' or 'Freedom of the Shannon') to make dolphinwatching a responsible, sustainable tourist attraction. Accredited operators agree to abide by a code of conduct overseen by the Shannon Dolphin and Wildlife Foundation. With the help of onboard naturalists, sightings are virtually guaranteed (trips lasting two to three hours have about a 95 per cent success rate) and the tours contribute to scientific monitoring and

Below: Short-beaked Common Dolphins (Delphinus delphis) are among several species visible from observation points on shore in south-western Ireland. From the underside, you can see this one's distinctive beak markings and white belly.

research. The dolphins are normally around from April until October, and spend most of their time in a 100 sq km (40 sq mile) area in the lower Shannon Estuary. Tours leave the old fishing village of Carrigaholt or the town of Kilrush, both on the north side of the estuary, and the dolphins can often be seen from local headlands such as Kilcredaun Head and Leck Point, in County Kerry. Short-beaked Common Dolphins and Minke Whales can sometimes be observed from Loop Head, overlooking the mouth of the Shannon.

Beyond Cape Clear

There are also whalewatching trips (usually combined with birdwatching) from Kinsale, Castletownhead, Baltimore, Schull and Clear Island, in Cork, in the south-western corner of the country. They explore the offshore waters beyond Mizen Head and Cape Clear, and frequently spot Harbour Porpoises, Bottlenose Dolphins, Short-beaked Common Dolphins, Risso's Dolphins, Long-finned Pilot Whales and Minke Whales. Killer Whales are occasionally seen, mainly in August and September (a pod spent several weeks in Cork harbour in 2001), and Atlantic White-sided Dolphins, White-beaked Dolphins and a wide variety of other species are encountered on a sporadic basis. Cuvier's Beaked Whales and Northern Bottlenose Whales have been recorded in this area, as well.

Most cetacean sightings in the region are from April to August, but in recent years there have also been visits from Sei and Fin Whales in the autumn and winter. Some five or six groups of Fin Whales are often seen between Seven Head and Galley Head (south-west of Cork); they seem to coincide with the arrival of big shoals of Sprat and Herring. There are also trips from Dingle and Kenmare, in neighbouring County Kerry, which explore the areas around Blasket Island and The Skelligs, and Kenmare Bay, respectively. Again, these trips tend to be general nature tours, observing everything from breeding seabirds to seals, but the whales, dolphins and porpoises are always an important feature. In particular, there are many Harbour Porpoises in the inner Kenmare Bay throughout the summer, and a population of Short-beaked Common Dolphins lives in the outer Bay area year-round.

Land-based whalewatching

The western and southern coasts of Ireland are widely regarded as offering some of the best land-based

*Opposite: Long-finned Pilot Whales (*Globicephala melas*) are a distinctive jet-black with a bulbous forehead. They are frequently encountered in the offshore waters beyond Mizen Head and Cape Clear Island.*

whalewatching in Europe. More than a dozen species can be seen along their rugged shorelines and from the many headlands facing the open Atlantic. Bolus Head (near Ballinskelligs) and Slea Head (on the end of the Dingle Peninsula) are particular favourites. There are even dedicated whale and dolphin walks; a real Irish innovation, these last anything up to a week and include pub lunches and bed and breakfast accommodation.

Northern Ireland

There is no whalewatching industry in Northern Ireland at the moment, but this is likely to change in the not-too-distant future. The region's long stretches of coastline, with scattered, rugged islands, are visited by a range of different cetaceans and there is already some wonderful, opportunistic whalewatching here.

Bottlenose Dolphins occur mainly off the north coast, Risso's Dolphins are seen off the north and north-west, and Harbour Porpoises are fairly common around the north and east coasts. Killer Whales turn up infrequently almost anywhere and there are very occasional sightings of Minke Whales and White-beaked Dolphins.

Shore-based whalewatching in Northern Ireland

The Irish Whale and Dolphin Group organizes timed research watches from a number of shore locations. They also collate any sightings reported by the public. Their research suggests a few likely hotspots for shore-based whalewatching. Harbour Porpoises are regularly seen along the stretch of coast from Castlerock (County Derry) to Ballintoy (County Antrim), especially in late summer and autumn; from Garron Point south to Island Magee (County Antrim); around Whitehead on the north coast of the lough mouth (County Antrim); from the Copeland Bird Observatory on Light House Island (6.4km (4 miles) from Donaghdee) in County Down; and around the entrance to Strangford Lough (County Down). Rathlin Island, 10km (6 miles) off the north coast facing Ballycastle, can be a good spot for Risso's Dolphins.

ITALY

Main species: Fin Whale, Sperm Whale, Cuvier's Beaked Whale, Long-finned Pilot Whale, Short-beaked Common Dolphin, Striped Dolphin, Bottlenose Dolphin, Risso's Dolphin.

Main locations: trips depart from San Remo, Imperia, Genoa, La Maddalena (Sardinia), Ischia Island.

Types of tours: half- and full-day tours, extended multi-day expeditions, research programmes, and land-based observation points.

When to go: most tours operate May to September, although many species are present year-round.

Contact details: Tethys Research Institute: www.tethys.org
StudioMare: www.mare.it/studiomare
Ecovolunteers: www.ecovolunteer.org
Santuario dei Cetacei: www.santuariodeicetacei.com or www.whale-watch.org
Riviera Line: www.rivieraline.it/whale.htm
BluWest: www.whalewatch.it/bluWesten.html

Italy has coasts on no fewer than four seas within the Mediterranean: the Ligurian, Tyrrhenian, Ionian and Adriatic. They are all good to outstanding for whale- and dolphinwatching. Most Italian-based tours currently focus on the Ligurian and Tyrrhenian seas, but the Italian Tethys Research Institute also runs the Ionian Dolphin Project, which is based in Greece, and established the Croatian-based Adriatic Dolphin Project (now run by a non-governmental organization called Blue World, in collaboration with the Croatian Natural History Museum).

Italian whalewatching

Whalewatching in Italy began relatively recently, in 1988, and is still fairly small-scale with just over 5,000 whalewatchers recorded each year in the late 1990s. A number of different vessels are used, including rigid-hulled inflatables, motor cruisers and liveaboard sailing schooners. StudioMare uses a beautiful sailing wooden cutter called the *Jean Gab*, which was built in 1930 and has been restored specifically for cetacean research (it even has towable hydrophones and an underwater video camera fixed on the bow). The tours are generally of a very high quality, and there is a strong

Tethys Research Institute

The Italian-based Tethys Research Institute is a non-profit organization dedicated to the study and conservation of the marine environment. Founded in 1986, it is particularly active in cetacean research and conservation within the Mediterranean. The main aim of its research is to help with developing effective management plans and achieving specific conservation goals. It also has an extensive public awareness and education programme, encouraging students and other volunteers to get involved in its work.

commitment to research, education and conservation. Several have been developed in partnership with research projects, which offer training in return for practical help with fieldwork. It is only a matter of time, however, before the industry develops and grows and, to ensure that it remains responsible, there is a determined effort to put regulations and codes of conduct in place before that happens.

The Ligurian Sea

The Ligurian Sea is the main region with commercial whalewatching. The ancient Romans were certainly aware of the local whales – they called the stretch of Italian coast between Imperia and Ventimiglia the 'Costa Balenae' – and recent research suggests that there are probably between two and four times as many cetaceans here than in most other parts of the Mediterranean (the Alborán Sea also has a high abundance). Enclosed by northern Italy, south-eastern France and northern Sardinia, the Ligurian is now regarded as a critical region for cetaceans.

The most sought-after whales in the area are Fin Whales, which are the only regular baleen whales in the Mediterranean. As many as 900 of them gather in the Ligurian Sea (out of a total Mediterranean population of about 3,500) from April to September (with a peak in July). Particular oceanographic features, including a permanent frontal system, provide ideal conditions for their main prey, a kind of shrimp-like krill, and they visit the Ligurian primarily to feed. They are genetically distinct from their contemporaries in the North Atlantic and are most likely to belong to a geographically isolated population, which lives in roughly the same area year-round. They are believed to make short migrations within the Mediterranean. The Ligurian is also home to some 25,000–40,000 dolphins during the summer. Striped Dolphins are by far the most common, but Risso's and Bottlenose Dolphins and Long-finned Pilot Whales also occur. Sperm Whales and Cuvier's Beaked Whales are found over deeper water here, too.

Ligurian Sea Cetacean Sanctuary

Conservation and research groups, concerned about the many threats to the region, campaigned throughout the 1990s for the Ligurian Sea to be given better protection. Finally, at a meeting held in Rome on 25 November 1999, the Ministers of the Environment from Italy, France and Monaco signed a treaty designating an area twice the size of

Switzerland (approximately 100,000 sq km (38,600 sq miles) as the Ligurian Sea Cetacean Sanctuary. The new sanctuary lies within an area marked by two lines drawn from Capo Falcone (north-western Sardinia) to Toulon (southern France), in the west, and from Capo Ferro (north-eastern Sardinia) to Fosso Chiarone (northern Italy), in the east.

It is the first time that several northern hemisphere countries have established a marine protected area in international waters. The treaty commits Italy, France and Monaco to coordinate research and monitoring activities in the Ligurian Sea and to intensify efforts to protect the region from pollution, destructive fishing methods and disturbance caused by high levels of boat and ship traffic. They also have to coordinate public awareness campaigns to inform locals and tourists alike about the sanctuary and its inhabitants.

Only time will tell how much impact sanctuary status will have on the cetaceans that make it their home. A major concern is that it is one of the most populated corners of the Mediterranean and Fin Whales feed in the area at the peak of the tourist season. With the Italian marine research institute, ICRAM, taking a leading role in preparing whalewatch regulations and policy for the sanctuary, there is real hope that whalewatching in the Ligurian Sea will continue to be responsible and of a high quality.

Ligurian Sea whalewatching

The Tethys Research Institute carries out research in the sanctuary under two main projects – both of which take paying volunteers for a week or more at a time. The first is the Mediterranean Fin Whale Project (MFWP), which was established in 1990 and focuses on the ecology and behaviour of Fin Whales in the pelagic zone between the French-Italian coast and Corsica. The second is the Squid Loving Odontocetes ProjEct (SLOPE), which was established in 1996 and studies the ecology of Cuvier's Beaked Whales, Risso's Dolphins, Bottlenose Dolphins, Striped Dolphins and other odontocetes over the continental slope area where the depth ranges from 200–2,000m (660–6,600ft). Volunteers are involved in all aspects of data collection, from photo-identification and audio recording to collecting faeces for diet analysis or skin samples for an investigation into genetic diversity and toxicology. A real effort is made to provide training in the field and through illustrated slide shows and lectures on shore.

Opposite: The waters around Capri are home to several different dolphin species throughout the summer.

Several commercial whalewatching trips are also available in the sanctuary, with a sightings success rate of over 90 per cent, and some of them have naturalists provided by Tethys.

Tyrrhenian Sea whalewatching

There are also some research projects in the Tyrrhenian Sea, which similarly rely on funding and assistance from volunteers. Two in particular – the Ischia Dolphin Project and the Ventotene Pilot Whale Project – are run by StudioMare, in collaboration with Tethys. Based on the island of Ischia, about 30km (19 miles) north-west of Capri and roughly opposite Naples, StudioMare hosts eight volunteers every week throughout the summer. The long-term aim of the Ischia Dolphin Project, which was set up in 1991, is to have Cuma Submarine Canyon included in a proposed marine protected area. The canyon is a critically important habitat, close to the northern coast of Ischia, and hosts large schools of Short-beaked Common, Striped, Bottlenose and Risso's Dolphins throughout the summer; it is believed to be an important feeding and breeding area, and newborn calves are seen regularly during July and August. The area is also an important feeding ground for large numbers of Fin Whales.

The Ventotene Pilot Whale Project is interesting because, outside the Alborán Sea and the Strait of Gibraltar, Long-finned Pilot Whales are rare in the Mediterranean. In 1995, however, researchers from StudioMare discovered a small pod of just six individuals off the south-west coast of the island of Ventotene (approximately 40km (25 miles) north-west of Ischia). They remain in a small area of about 3 sq km (1¼ sq miles) every summer, and have been studied by StudioMare ever since.

There is also a research project to study and protect resident Bottlenose Dolphins around the small island of La Maddalena, off the north-east coast of Sardinia. Strictly speaking, the study area is located in the northern Tyrrhenian Sea, but it is included in the southern portion of the Ligurian Sea Cetacean Sanctuary. The aim is to develop conservation measures for these inshore dolphins, which are threatened by coastal activities such as fishing, tourism and pollution. The work is being done by Ecovolunteers and, after a short training period, participants are able to help with photo-identification and behavioural studies conducted from a small inflatable boat and from an observation point on shore.

NORWAY

Main species: *northern Norway:* Minke Whale, Sperm Whale, Killer Whale, Long-finned Pilot Whale, Harbour Porpoise; *Svalbard:* Beluga.

Main locations: Andenes, Stø, Tysfjord, Svalbard.

Types of tours: half- and full-day tours, extended multi-day expeditions, research programmes and land-based observation points; snorkelling and diving with Killer Whales in Tysfjord.

When to go: *Andenes:* late May to mid-September; *Stø:* 1 June to 31 August; *Tysfjord:* mid-October to mid-January (best daylight mid-October to mid-November but more whales mid-November to mid-January); *Svalbard:* late May to late August (depending on ice break-up).

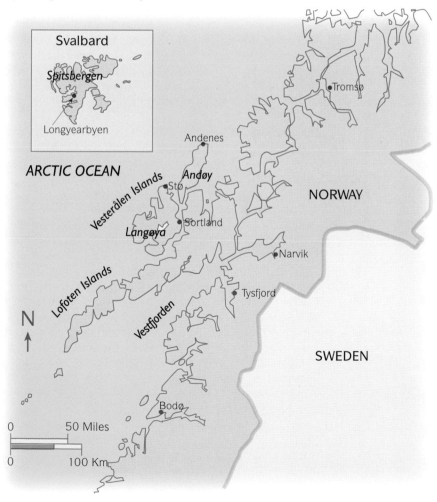

Contact details: *Andenes*: Whale Safari Ltd (Hvalsafari AS): www.whalesafari.com
Whale Tours Ltd: www.whaletours.no
Tysfjord: Tysfjord Turistsenter AS: www.tysfjord-turistsenter.no (safari details: www.orca-tysfjord.nu)
Lofoten Opplevelser: www.lofoten-opplevelser.no
Svalbard: Svalbard Wildlife Services: www.wildlife.no
Origo Expedition: www.origoexpedition.se

Outside Norway: Quark Expeditions: www.quarkexpeditions.com
Arcturus Expeditions: www.arcturusexpeditions.co.uk
Discover the World: www.arctic-discover.co.uk

Norway's first whalewatching trip set off from the rugged northern coast in 1988, to look for Sperm Whales some 300km (185 miles) above the Arctic Circle. The trip was an enormous success and prompted the development of one of the highest quality whalewatch industries in the world. In fact, Whale Safari Ltd – the largest and first operator in the country – is now widely admired for providing the ultimate combination of education, research and tourism on its tours.

There are two premier whalewatch destinations on mainland Norway: the Lofoten–Vesterålen islands, just an hour's sail from a major feeding ground used by Sperm Whales every summer, and nearby Tysfjord, which is a major feeding ground for Killer Whales every winter. There are also some whalewatching opportunities around the Arctic archipelago of Svalbard.

Tysfjord

Tysfjord is sometimes described as Norway at its deepest and narrowest. This world of moody Norwegian fjords is a spectacular destination in its own right – but has the added advantage of being one of the easiest and most reliable places for watching Killer Whales. Sometimes, Long-finned Pilot Whales have also been seen in recent years.

Killer Whales have not always been coming here and began visiting only when their preferred prey, Herring, moved into the area. During the 1980s, there was a sudden change in the migratory habits of Herring when, quite suddenly, they began to move into coastal waters for the winter. One theory is that the change was a direct result of over-fishing in the north-east Atlantic but, whatever the reason, in

1987, almost overnight, Tysfjord became the winter home of a large proportion of the North Atlantic Herring population. With the Herring came the Killer Whales – and both the fish and their predators have been returning every year since. Research on Tysfjord's newcomers began two years later, in 1989, and has continued ever since. Whalewatch operators have cooperated with researchers since the beginning and support their work financially. There are believed to be as many as 600 Killer Whales in the area each winter and they spend much of their time feeding.

Unfortunately, this is all happening at a time when daylight is in short supply. The Killer Whales are normally in the area

Below: *Whalewatching in Tysfjord, northern Norway, offers the chance to see Killer Whales (Orcinus orca) against a spectacular mountainous backdrop.*

from mid-October until mid-January and Tysfjord is so far north that there are long periods of darkness for much of that time. The prime viewing season for both whales and light is very short – about six to eight weeks from October to early December – and, even then, during the darkest period, whalewatching trips can be run for only four to six hours during the middle of the day. However, whalewatchers are invited to attend an informative and entertaining lecture and slide show in the half-light before the trip and, if the sky is clear, there is a good chance of seeing the spectacular northern lights after returning to shore.

The whales are often close to land and can be seen from various vantage points on the outskirts of town, but they are best seen on organized boat tours. They are nearly always found within an hour or two of leaving the dock and it is not surprising that whalewatching here has become big business. Its popularity caused so many traffic jams on the local roads and in the fjords in the past that operators and researchers were forced to improve the situation. However, the whalewatching is good for the local hotels, adding much-needed off-season income, and they have reciprocated by organizing a variety of educational workshops and events.

Andenes

Commercial whalewatching first began in Norway a little farther north, beyond Tysfjord, in the colourful fishing port of Andenes on the northern tip of Andøy Island. In 1986, a group of enthusiastic and like-minded biologists hatched a plan to find a suitable area for both whalewatching and whale research. They hit upon the Lofoten-Vesterålen group of islands as a likely starting point and set off in a

Carousel feeding

Tysfjord's Killer Whales use a rather unusual cooperative hunting technique called carousel feeding, in which they herd shoals of Herring against the surface of the sea. They swim around in ever-decreasing circles, sometimes blowing bubbles, and force the fish into a tight ball before stunning them with their powerful tails and then picking them out of the sea one by one.

small converted fishing boat to look for whales. At a time when most Norwegians saw whales as a resource only when they were dead, the aim of the project was to demonstrate that living whales could be a valuable resource, too.

The biologists were hoping to find an area with a reliable population of whales, which was also close enough to land to make day trips possible – and they succeeded. Where the edge of the continental shelf comes very close to Andenes, just 20–30km (12–19 miles) offshore, the ocean basin forms a submarine canyon. This 1,000m (3,300ft) canyon, they discovered, is a prime feeding area for Sperm Whales.

The following year, the dedicated team of professional whalewatchers invited people from the tourist industry, business and local government, as well as journalists and fellow biologists, to join them on a special trip. Many of these people were quite sceptical but, when the big day arrived, they seemed to be surrounded by Sperm Whales and even had a playful school of White-beaked Dolphins riding the bow wave of one of the boats. As if that was not enough to convince everyone, Al Alvelar, an ex-fisherman from Cape Cod, in New England, came over to Norway to tell everyone how he became a whalewatch captain – and a millionaire.

Gradually, the scheme took shape. Even some Norwegian whalers began to take a serious interest in exploiting this new source of income, using all their skills and experience gained after long years on the whaling boats. Nowadays, whalewatching seems to dominate life in Andenes. Wandering around town, you see whales everywhere – in the supermarket, the bank and the post office, as well as on posters, postcards and T-shirts. Whalewatching here is probably the most culturally diverse in the world. Every year, visitors from 40 different countries travel to Andøy Island (which is connected to the mainland by bridges) specifically to watch the Sperm Whales.

Unlike Tysfjord, Andenes does not have a problem with daylight. The whalewatching here takes place in summer and, indeed, the region enjoys 24-hour daylight for a couple of months during May, June and July. Whale Safari Ltd offers educational trips, led by multi-lingual researchers and students, lasting four to five hours. It normally takes about an hour to reach the whales, which

Below: *There are several good places to see Belugas (*Delphinapterus leucas*) around the remote archipelago of Svalbard. These animals generally occur in groups. Not all Belugas are white, they become so when they are sexually mature at 5–10 years old.*

feed in an area called Bleiksdjupet, north-west of Andøy Island, and this leaves plenty of time to watch them properly. There is a 95 per cent success rate – but if you are really unlucky and do not see any whales you get a free trip another day. Two boats are used, with a carrying capacity of 99 and 80 passengers respectively, and there are trips several times a day, as required. In mid-summer, there are even late evening departures to watch the whales bathed in the light of the midnight sun.

Several other cetaceans can be encountered on these trips, besides the Sperm Whales. Minke Whales are seen fairly frequently and Long-finned Pilot Whales have been

observed quite often in recent years. With a lot of luck, it is also possible to see Fin Whales, Humpback Whales, Killer Whales and White-beaked Dolphins. Harbour Porpoises are fairly common in summer.

A free visit to the Andenes Whale Centre is included in the cost of a whalewatching ticket. Established in the late 1980s, in an old fish-processing factory on the harbourside, this is where visitors can learn about whales, whale research and north Norwegian whaling. There are public lectures on whale behaviour, ecology and other topics, an audio visual show and a well-stocked library.

Mature male Sperm Whales

Most of the Sperm Whales recorded near Andenes are presumed to be mature males and they visit northern Norway to feed on deep-water squid. More than half of them are seen only once during the summer and then never return. Others have been seen on numerous occasions, but only during one summer. There are a few others that seem to be long-term residents.

Andenes is the most northerly place in the world to see Sperm Whales in large numbers and most of our knowledge about this particular population comes from research conducted during the tourist trips. The onboard naturalists provide informative commentaries, and conduct photo-identification work and other studies from the whalewatch boats. In particular, they have been working on a Sperm Whale catalogue, which contains photographs and information for every individual whale they have ever encountered. During the period 1987–2000, they recorded no fewer than 2,900 sightings and there are now more than 370 different Sperm Whales in the catalogue. Pictures of every whale, together with all their personal details, are available online at http://130.241.163.46/spermwhale/index.html.

One smaller whalewatch operator visits the same area: Whale Tours Ltd, which is based in Stø on the northern tip of Langøya Island, south-west of Andøy. They take visitors to see the Sperm Whales, which are some 40km (25 miles) away, from the beginning of June to the end of August (and also encounter other species seen on Andenes trips). The tours last five to seven hours and, if no whales are seen, they also offer a free trip. Like Whale Safari Ltd, in Andenes, they take photo-identification pictures for research.

Svalbard

Much farther north there is limited whalewatching in the high Arctic – in Svalbard. This beautiful archipelago is one of the northernmost inhabited areas in the world. Various baleen whales feed in the surrounding seas during the summer, particularly in the comparatively ice-free western and southern parts, but Belugas are the real attraction. They move very close to shore and even into river mouths from May onwards and remain until the end of August or early September (precise timings depend on ice break-up in the fjords). As they chase Polar Cod and Capelin into the shallows, some even get stuck on the mud at low tide. There are several notable hotspots for watching Belugas from shore: along the spit south of the airport at Sveagruva (where the water is normally clear); on the western side of the Adventdalen river mouth, outside Longyearbyen (where the water tends to be muddy and so it is only possible to see the whales as they surface); and the Bellsund-Van Mijenfjorden-Van Keulenfjorden areas of the west coast of Spitzbergen.

It is possible to fly to Longyearbyen (during the summer there are almost daily flights from Tromsø, on mainland Norway) and to join excursions in small boats. The easiest way to whalewatch here, however, is to join an expedition cruise ship. There are several two- or three-week cruises to choose from and, although they are not dedicated whalewatching trips, many of them focus on natural history and whales are an integral feature. Rigid-hulled inflatables are sometimes used for exciting close encounters with particularly cooperative whales. Some of these trips start in Longyearbyen, on vessels that are based in the region, but the bigger expedition cruises tend to include Svalbard on longer trips from Aberdeen, Reykjavík and other ports farther south.

Hunting whales

With so much interest in whales, and such a successful whalewatching industry, it is perhaps surprising that Norway still hunts whales. When the International Whaling Commission (IWC) voted for an indefinite moratorium on commercial whaling to come into effect from the mid-1980s, Norway officially objected. Under IWC rules, it has been legally entitled to continue hunting whales ever since. Its annual quota (set by the Norwegian government) is steadily increasing and, in 2001, no fewer than 550 Minke Whales were killed. With no end to the hunt in sight, maybe the best hope in the meantime is for commercial whalewatching and wild whale research at least to contribute to a better understanding of the true value of living whales.

PORTUGAL

Main species: Bottlenose Dolphin.

Main location: Sado River estuary.

Types of tours: half-day tours, research programmes and land-based observation points.

When to go: tours operate year-round.

Contact details: Vertigem Azul: www.vertigemazul.com

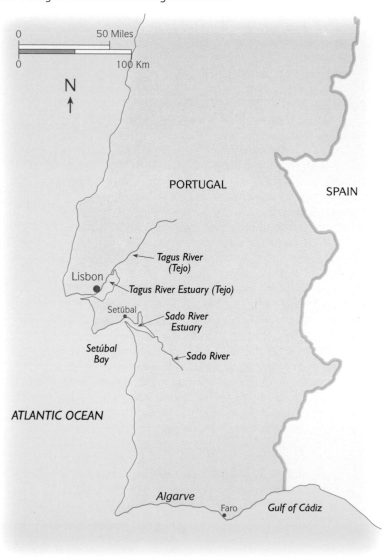

Until the 1960s, there were two resident populations of Bottlenose Dolphins in Portugal: one in the estuary of the Sado River, 48km (30 miles) south of Lisbon, and the other in the Tagus River. The population in the Tagus declined and eventually disappeared altogether, apparently unable to cope with a barrage of threats ranging from pollution and increasingly heavy boat traffic to a diminishing food supply due to overfishing.

Facing similar threats, the Sado population has also been struggling in recent years. There were around 50 individuals in the 1980s but a small resident population of only 32 survive today. They are unique in Portugal because they remain in one area year-round. A study in 1997 found that many of the dolphins have skin disorders, which suggests that they may have depressed immune systems, and there is now concern that this small and ageing population could experience reproductive problems in the future.

Dolphinwatching began in Portugal in the early 1980s although, until 1998, it was primarily land-based. There is little opportunity for land-based observations now, with such a small population, but there are organized boat trips. A rigid-hulled inflatable is used for two-and-a-half to three-hour tours, which leave from Setúbal and explore the Sado Estuary Natural Reserve (set up as a marine protected area partly in recognition of the dolphins) between the mountains of Arrábida Natural Park and the Atlantic Ocean.

The tours begin with an educational slide presentation to introduce the dolphins, as well as the ecology and history of the area. In addition, the 6m (20ft) boat is equipped with a hydrophone to listen to the dolphins vocalizing underwater (although this is not always possible if there is too much noise from nearby boats and jetskis). A long-standing photo-identification project enables researchers (and dolphinwatchers) to recognize particular individuals in the population, using their scars and other distinctive markings. There is a 96 per cent success rate on these tours, so sightings are virtually guaranteed.

There may be considerable potential for whalewatching in other parts of mainland Portugal (it is well established in the Azores) but it has yet to be developed. The coast of the Algarve and the Gulf of Cádiz, both in the southern part of the country, are considered particularly promising.

SPAIN

Main species: Fin Whale, Sperm Whale, Long-finned Pilot Whale, Killer Whale, Short-beaked Common Dolphin, Striped Dolphin, Bottlenose Dolphin.

Main locations: Algeciras Bay, Strait of Gibraltar, Alborán Sea, Bay of Biscay (see separate entry).

Types of tours: half- and full-day tours, extended multi-day expeditions, research programmes and land-based observation points.

When to go: most tours in the south are offered year-round (some operators close for a couple of months); dolphins in Algeciras Bay present year-round; Long-finned Pilot Whales and Fin Whales are year-round; Sperm Whales usually in the spring and summer; Killer Whales mainly July, August and early September.

Contact details: Firmm España: www.firmm.org
Mar de Ballenas: www.mardeballenas.com
Alnitak: www.geocities.com/tofte2000
Whale Watch España: www.whalewatchtarifa.com
Spanish Cetacean Society: www.cetaceos.com
CIRCE: www.circe-asso.org
Turmares Tarifa S.L.: www.windtarifa.com/turmares
Arion: www.turmares.com
Birskas Nautic Sport: www.marbella2000.com/briskas

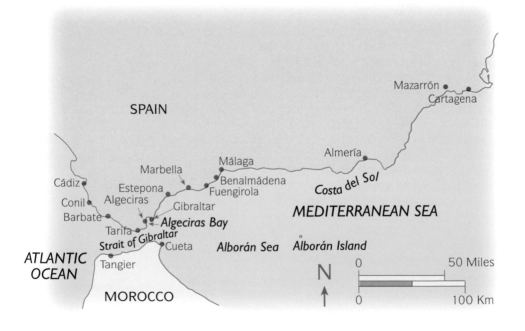

Whale- and dolphinwatching is relatively new to Spain and did not really begin until the early 1990s. It has developed rapidly in the past decade, however, especially in the Costa del Sol region along the southern coast of Andalucía, and now attracts more than 40,000 people every year.

In the south of the country, whalewatch trips leave the ports of Tarifa, Algeciras, Estepona, Marbella, Fuengirola, Benalmádena, Malága, Almería and Mazarrón year-round. There are also trips from La Línea (see Gibraltar entry). Between them, they take whalewatchers to three main regions: Algeciras Bay, the Strait of Gibraltar and the Alborán Sea. Using a wide variety of vessels, from kayaks and rigid-hulled inflatables to small fishing boats, glass-bottomed boats and speedboats, they offer an equally wide variety of tours. Most last for a few hours, or a day, but others are extended multi-day voyages or research expeditions. There are also one- and two-week whalewatching courses, combining theory lessons with daily boat trips, which are run by Firmm España (the Foundation for Information and Research on Marine Mammals) from their centre in Tarifa. There are some good land-based observation points for dolphinwatching around Algeciras Bay. In Spain, the main ones are Punta del Carnero, near the abandoned whaling station on the western side, and next to the oil refinery at the northern end of the Bay (see separate Gibraltar entry).

There is also some good whale- and dolphinwatching from Ceuta, a small Spanish enclave on the north coast of Africa, and there are some excellent vantage points to watch from shore. Ceuta is also a centre for aerial whalewatching – the three-hour flights search for Killer Whales, Sperm Whales and Fin whales and offer one of the few opportunities in Europe to watch whales from the air.

Many of Spain's tours are both responsible and informative and, indeed, some are run by non-governmental research and conservation organizations such as Alnitak. In Andalucía, the boats have as many as three naturalists on board and the whalewatch companies in Tarifa have agreed and signed an ethical code (meanwhile, the Spanish government is currently drafting official regulations for whalewatching). Several whalewatch companies and non-governmental organizations in the region are also running educational programmes and conducting extensive research.

Algeciras Bay and the Strait of Gibraltar

Algeciras Bay is home to Short-beaked Common and Striped Dolphins, which are present in abundance throughout the year, and Bottlenose Dolphins are seen less frequently. There is an average sightings success rate of 95–100 per cent within the Bay.

The Strait of Gibraltar separates the southernmost point of Spain from the African coast and is just 16km (10 miles) wide at its narrowest point. This is where the Atlantic Ocean meets the Mediterranean Sea and it is an area with a rich diversity of cetacean species. As well as the three local dolphins, it is possible to see Fin Whales, Sperm Whales, Long-finned Pilot Whales and Killer Whales with some regularity a little farther offshore. Fin Whales pass through the Strait in small numbers throughout the year, Sperm Whales are seen mainly in the spring and summer, and Long-finned Pilot Whales are resident in the centre of the Strait. The Killer Whales are a draw for many whalewatchers – they spend a couple of months here each summer to feed on tuna and are seen mainly in July, August and early September. They are sometimes very close to shore, but can be some distance out into the Strait. Cuvier's Beaked Whales and Risso's Dolphins are also seen from time to time and several other species have been recorded over the years.

The Alborán Sea

The Alborán Sea offers a similar range of species, but without the Killer Whales. Striped and Short-beaked Common Dolphins are seen frequently, along with smaller numbers of Bottlenose Dolphins. Long-finned Pilot Whales are resident within a few kilometres of shore and readily associate with whalewatch boats from Estepona, Marbella and Benalmádena, while several other species are seen occasionally. This biologically-rich region in the western Mediterranean was named after a tiny Spanish island lying roughly midway between Almería and Melilla, and is bordered by mainland Spain, the north coast of Africa and the Balearic Islands. One of the best ways to explore the Alborán Sea is to join a research trip run by the Alnitak Marine Environment Research and Education Centre. Every nine days groups of eight Earthwatch volunteers board their 18m (60ft) wooden ketch (a converted 1910 Norwegian fishing vessel called the *Toftevaag*) to help with cetacean surveys. No particular skills are required, but people with previous marine mammal research experience or Zodiac driving are preferred. One of

Opposite: *Striped Dolphins (*Stenella coeruleoalba*), which seem to spend an inordinate amount of their time in the air, are frequently seen in both Algeciras Bay and the Alborán Sea.*

the aims of the study is to find out why Short-beaked Common Dolphins have been declining in the region and to put forward proposals for their conservation. These are genuine research trips, sometimes working in areas with very low cetacean densities, and may be frustrating for anyone more used to tourist-style whalewatching.

Bay of Biscay

There are also trips to watch Bottlenose Dolphins off the north-west coast of Spain, in the Galicia region, and some of the best whalewatching in the whole of Europe takes place in the Bay of Biscay. More than a dozen species are seen regularly from two passenger cruise-ferries plying the Bay between southern England and northern Spain (this outstanding region is dealt with in the Bay of Biscay entry).

Below: *A playful Bottlenose Dolphin* (Tursiops truncatus) *frolics in the wake of a whalewatching boat.*

WESTERN ARCTIC RUSSIA

Main species: Bowhead Whale, Narwhal, Beluga.

Main locations: White Sea (Beloye More), Taymyr Peninsula, Franz Josef Land, Novaya Zemlya, Severnaya Zemlya.

Types of tours: half- and full-day tours, extended multi-day expeditions, research programmes and land-based observation points.

When to go: summer (late June to August best).

Contact details: Ecological Travels Centre: www.ecotravel.ru/eng/main.html
Outside Russia: Arcturus Expeditions: www.arcturusexpeditions.co.uk
Quark Expeditions: www.quarkexpeditions.com
Ecovolunteer: www.ecovolunteer.org

There are relatively few organized whalewatching tours in the Russian Arctic, mainly because the logistics can be extremely difficult, but the region is opening up to ecotourism and the potential is quite considerable. There are already some great opportunities for watching the three Arctic whale species (Bowhead, Narwhal and Beluga), with the added bonus of Polar Bears, Arctic Foxes, Walruses and a fabulous variety of other wildlife.

Expedition cruise ships

The easiest way to whalewatch here is to join an expedition cruise ship, such as the ice-breaker *Yamal*. This is a Russian ship with a Russian crew, although the expedition teams are mostly American or western European and the tours tend to be operated by companies based in the USA or Britain. The *Yamal* is one of several ships that spend the northern summer in the Arctic and the southern summer in the Antarctic (others include the *Professor Molchanov* and *Kapitan Khlebnikov*). These two- or three-week cruises are not dedicated whalewatching trips, but many of them focus on natural history and whales are an integral feature. Rigid-hulled inflatables are used for exciting close encounters and, if weather conditions allow, the ice-breakers also use helicopters to find the whales and sometimes allow passengers to view them from the air. The itineraries inevitably vary from company to company and from year to year, but the Taymyr Peninsula is often included along with one or two of the high Arctic archipelagos of Franz Josef Land, Novaya Zemlya and Severnaya Zemlya. Most of these cruises start in Murmansk, Russia, or Longyearbyen on Svalbard, Norway, and are mainly in July and August.

Below: *With their long, spiralling tusks, male Narwhals (*Monodon monoceros*) are the whales every whalewatcher wants to see on a visit to Arctic Russia.*

Belugas

The White Sea is particularly good for Belugas and it has been estimated that there are some 800 of them (give or take 100) in the southern portion. For two months every summer, large numbers of Belugas congregate to mate and calve in shallow water around the Solovetsky archipelago. This beautiful group of islands lies in the boreal region of northern European Russia, but has 20 hours of daylight every day in the summer and the marine environment here is considered sub-arctic. There has been a research project, since 1994, to study the social behaviour and communication of the whales. Operated by the Marine Bioacoustic Laboratory of the Shirshov Institute of Oceanology, which is part of the Russian Academy of Science, it is open to volunteers who can join the research team for two weeks at a time. Everyone camps near the shore and helps to collect food such as fish, berries and mushrooms nearby; food is cooked on open fires built from driftwood. A three-platformed observation tower has been built on the coast of a small island, near Beluzhy Cape, and can be reached on foot at low tide. This is the best time and place to view the whales and some 2–40 can be observed almost daily from the end of June to mid-August. The volunteers use hydrophones and video cameras to record vocalizations and social behaviour.

Belugas can also be observed on boat trips from shore around the Solovetsky archipelago and off the Taymyr Peninsula on the north coast of Siberia. Every summer, large numbers of these snowy-white whales move into Arctic river mouths in this region (the northernmost point on mainland Siberia) and the estuaries of the Yeniseyskiy and Khatangskiy gulfs, on either side of the Taymyr, are particularly good.

New regulations

Unfortunately, in summer 2001, there were complaints about irresponsible boat operators disturbing the Belugas around the Solovetsky archipelago and, at the time of writing, the Russian authorities have just introduced new regulations to stop anyone approaching the whales until further notice.

FURTHER READING

Carwardine, Mark, *Eyewitness Handbooks: Whales, Dolphins and Porpoises*, Dorling Kindersley, 1995.

Carwardine, Mark, Hoyt, Erich, Fordyce, R. Ewan, and Gill, Peter, *Whales & Dolphins: The Ultimate Guide to Marine Mammals*, HarperCollins, 1998.

Clapham, Phillip J., Powell, James A., Reeves, Randall R., and Stewart, Brent S., illustrated by Folkens, Peter A., *Sea Mammals of the World*, A&C Black, 2002.

Darling, James D., Nicklin, Charles 'Flip', Norris, Kenneth S., Whitehead, Hal, and Würsig, Bernd, *Whales, Dolphins and Porpoises*, National Geographic Society, Washington, 1995.

Hoyt, Erich, *The Whale Watcher's Handbook*, Doubleday, 1984.

Hoyt, Erich, *Meeting the Whales The Equinox Guide to Giants of the Deep*, Camden House, 1991.

Hoyt, Erich, *Riding with the Dolphins The Equinox Guide to Dolphins and Porpoises*, Camden House, 1992.

Evans, P.G.H., *The Natural History of Whales and Dolphins*, Christopher Helm, 1987.

Katona, Steven and Lien, Jon, *A Guide to the Photographic Identification of Individual Whales*, American Cetacean Society, 1990.

Leatherwood, Stephen, and Reeves, Randall R., *The Sierra Club Handbook of Whales and Dolphins*, Sierra Club Books, 1983.

Martin, Anthony R., *Whales and Dolphins,* Salamander Books, 1990.

Simmonds, Mark P. and Hutchinson, Judith D. (eds), *The Conservation of Whales and Dolphins: Science and Practice*, J. Wiley, 1996.

Waller, Geoffrey (ed.), *Sealife: A Complete Guide to the Marine Environment*, Pica Press, 1996.

WorldLife Library (various authors), *Blue Whales; Minke Whales; Humpback Whales; Grey Whales; Sperm Whales; Killer Whales; Beluga Whales; Bottlenose Dolphins; Porpoises*; Colin Baxter, 1994-2000.

GLOSSARY

Amphipod small, shrimp-like crustacean that is a food source for some whales.

Baleen/baleen plates comb-like plates hanging down from the upper jaws of most large whales; used instead of teeth to capture prey.

Baleen whale sub-order of whales with baleen plates instead of teeth, known in the scientific world as Mysticeti.

Beak elongated snout of many cetaceans.

Blow cloud of water vapour exhaled by cetaceans (also known as the 'spout'); often used to describe the act of breathing.

Blowhole(s) nasal opening(s) or nostril(s) on the top of the head.

Blubber layer of fat just beneath the skin of marine mammals; important for insulation instead of fur.

Bow-riding riding in the pressure wave in front of a boat, ship or large whale.

Breaching leaping completely (or almost completely) out of the water, and landing back with a splash.

Bubble-netting feeding technique used by Humpback Whales in which they produce fishing nets by blowing bubbles underwater.

Callosity area of roughened skin on the head of a right whale, to which whale lice and barnacles attach.

Cetacean any member of the order Cetacea, which includes all whales, dolphins and porpoises.

Copepod a small crustacean that is a food source for some whales.

Crustacean member of a class of invertebrates (animals without backbones) that are food for many marine animals; mostly aquatic.

Dorsal fin raised structure on the back of most (but not all) cetaceans; not supported by bone.

Drift net fishing net that hangs in the water, unseen and undetectable, and is carried freely with the ocean currents and winds; strongly criticized for catching everything in its path, from seabirds and turtles to whales and dolphins.

Echolocation process of sending out sounds and interpreting the returning echoes to build up a 'sound picture', as in sonar; used by many cetaceans to orientate, navigate and find food.

Flipper flattened, paddle-shaped limb of a marine mammal; refers to the front limb of a cetacean (also known as the 'pectoral fin').

Flipper-slapping raising a flipper out of the water and slapping it onto the surface.

Fluke horizontally flattened tail of a cetacean; contains no bone.

Fluking raising the tail flukes into the air upon diving.

Gill net similar to a drift net in design, although much smaller and fixed in one position near the coast or in a river.

Hydrophone waterproofed, underwater microphone.

Glossary

Keel distinctive bulge on the tailstock near the flukes.

Krill small, shrimp-like crustaceans that form the major food of many large whales; there are about 80 different species, ranging from 8–60mm in length.

Lobtailing slapping tail flukes against the water, creating a splash.

Logging lying still at or near the surface.

Mandible lower jaw of the skull.

Melon fatty organ in the bulging forehead of many toothed cetaceans, believed to be used to focus sounds for echolocation.

Pectoral fin see 'flipper'.

Photo-identification technique for studying cetaceans using photographs as a permanent record of identifiable individuals.

Pod coordinated group of whales; term normally used for larger, toothed whales.

Polar region around either the North Pole or South Pole (i.e. Arctic or Antarctic).

Porpoising leaping out of the water while swimming at speed.

Purse-seine net long net set around a shoal of fish, then gathered at the bottom and drawn in to form a 'purse'.

Rorqual baleen whale of the genus *Balaenoptera*; many experts also include the Humpback Whale (genus *Megaptera*) in this group.

Rostrum upper jaw of the skull (may be used to refer to the beak or snout).

School coordinated group of cetaceans; term normally used in association with dolphins.

Snout see 'beak'.

Sonar see 'echolocation'.

Spout see 'blow'.

Spyhopping raising the head vertically out of the water, apparently to look around above the surface.

Submarine canyon deep, steep-sided valley in the continental shelf.

Tailstock muscular region of the tail between the flukes and the dorsal fin.

Temperate mid-latitude regions of the world between the tropics and the poles.

Toothed whale sub-order of whales with teeth, known in the scientific world as Odontoceti.

Tropical low-latitude regions of the world between the tropics of Capricorn and Cancer.

Tubercle circular bumps found on some cetaceans; usually along the edges of flippers and dorsal fins, but also on a Humpback Whale's head.

Turbid term used to describe muddy or cloudy water carrying lots of sediment.

Wake-riding swimming in the frothy wake of a boat or ship.

Whalebone another name for baleen.

Index

186

Whale Sighting Sheet

Date / time	Species	Lat	Lon	

	Place	Wind / sea	Est. number	Notes

Whale Sighting Sheet

Date / time	Species	Lat	Lon	

	Place	Wind / sea	Est. number	Notes

ACKNOWLEDGEMENTS

Many people have put a lot of time and energy into this book and it could not have been written without their help. I am grateful to them all for their wonderful generosity.

In particular, I would like to thank all the whalewatch operators listed in the *Where to Watch Whales in Europe* section of the book. They willingly answered my numerous e-mail and telephone enquiries and then checked the final text. It has been a great pleasure corresponding with them and I admire the work they are doing.

I would like to pay particular tribute to my great friend and co-conspirator, Erich Hoyt, who acted as a consultant on the book. It has benefited enormously from his extensive knowledge of the whalewatch industry. I was also very fortunate to have the help of three diligent and enthusiastic researchers – Caroline Williams, Christine Bathgate and Lisa Browning, who were brilliant as always.

Special thanks go to Jo Hemmings and Lorna Sharrock at New Holland Publishers, for their never-ending support, encouragement, enthusiasm and patience – even when the book took much longer to research and write than any of us had anticipated.

None of these people are in any way responsible for any errors of fact or emphasis that remain. I hope these are few and far between, but the credit for them goes to me.

Mark Carwardine

Publisher's Acknowledgements
All photographs by Mark Carwardine, with the exception of the following:
Ardea: Francois Gohier: p154; Doc White: pp136, 169, 180.
Bruce Coleman Collection: Johnny Johnson: boat on cover.
Sylvia Cordaiy Photo Library Ltd: Jill Swainson: p160.
Oxford Scientific Films: Tony Bomford: p124; Godfrey Merlen: p77.
Still Pictures: Roland Seitre: pp69, 177.
Dylan Walker: pp60, 81, 96, 116, 120.

All artwork by Martin Camm.

Stop Press: Please note that at time of going to press Finland now has a small Beluga-watching industry that may well develop further.
Contact details: Kon-Tiki Tours: www.kontiki.fi